||||| ||| || ||||| ||| ||||| |||||
I0134575

TC 3-04.35

Training Circular
No. 3-04.35

Headquarters
Department of the Army
Washington, DC, 1 March 2013

AIRCREW TRAINING MANUAL, UTILITY HELICOPTER, Mi-17 SERIES

Contents

			Page
	PREFACE		vii
Chapter 1	INTRODUCTION		1-1
	1-1. Crew Station Designation		1-1
	1-2. Symbol Usage and Word Distinctions		1-1
Chapter 2	TRAINING		2-1
	2-1. Qualification Training		2-1
	2-2. Refresher Training		2-1
	2-3. Mission Training		2-2
	2-4. Continuation Training		2-5
	2-5. Task List		2-10
	2-6. Currency Requirements		2-11
	2-7. Chemical, Biological, Radiological, Nuclear, and High Yield Explosive Training		2-11
Chapter 3	EVALUATION		3-1
	3-1. Evaluation Principles		3-1
	3-2. Grading Considerations		3-2
	3-3. Crewmember Evaluation		3-2
	3-4. Evaluation Sequence		3-3
	3-5. Additional Evaluations		3-7
Chapter 4	CREWMEMBER TASKS		4-1
	4-1. Task Contents		4-1
	4-2. Tasks		4-5
Chapter 5	MAINTENANCE TEST PILOT TASKS		5-1
	5-1. Task Contents		5-1
	5-2. Task List		5-2
Chapter 6	AIRCREW COORDINATION		6-1

*This publication supersedes TC 3-04.35, dated 24 March 2010.

6-1. Aircrew Coordination Background and Planning Strategy 6-1

6-2. Aircrew Coordination Principles 6-1

6-3. Aircrew Coordination Objectives 6-4

6-4. Standard Crew Terminology 6-4

Appendix A   NONRATED CREWMEMBER TRAINING AND QUALIFICATION Appendix-1

GLOSSARY Glossary-1

REFERENCES References-1

INDEX Index-1

Tasks

Task 1000 - Participate in a Crew Mission Briefing 4-6

Task 1002 - Conduct Passenger Briefing 4-9

Task 1004 - Plan a Visual Flight Rules Flight 4-10

Task 1006 - Plan an Instrument Flight Rules Flight......................................4-12

Task 1010 - Prepare a Performance Planning Card......................................4-14

Task 1012 - Verify Aircraft Weight and Balance 4-19

Task 1013 - Operate Mission Planning System..4-20

Task 1014 - Operate Aviation Life Support Equipment 4-21

Task 1016 - Perform Internal Load Operations..4-22

Task 1019 - Perform Preventive Maintenance Daily Check 4-24

Task 1020 - Prepare Aircraft for Mission 4-25

Task 1022 - Perform Pre-Flight Inspection 4-26

Task 1024 - Perform Before-Starting Engine through Before-Leaving
   Helicopter Checks..4-27

Task 1026 - Maintain Airspace Surveillance...4-29

Task 1028 - Perform Hover/Power Check 4-31

Task 1032 - Perform Radio Communication Procedures 4-33

Task 1034 - Perform Ground Taxi 4-35

Task 1038 - Perform Hovering Flight 4-37

Task 1040 - Perform Visual Meteorological Conditions Takeoff....................4-40

Task 1044 - Navigate by Pilotage and Dead Reckoning 4-43

Task 1046 - Perform Electronically Aided Navigation...................................4-44

Task 1048 - Perform Fuel Management Procedures......................................4-45

Task 1052 - Perform Visual Meteorological Conditions Flight Maneuvers....4-47

Task 1054 - Select Landing Zone/Pickup Zone/Holding Area 4-49

Task 1058 - Perform Visual Meteorological Conditions Approach 4-51

Task 1062 - Perform Slope Operations 4-54

Task 1064 - Perform Roll-On Landing 4-56

Task 1068 - Perform Go-Around.. 4-58

Task 1070 - Respond to Emergencies.. 4-60

Task 1074 - Respond to Engine Failure at Cruise Flight... 4-62

Task 1075 - Perform Single-Engine Landing... 4-63

Task 1082 - Perform Autorotation.. 4-64

Task 1094 - Perform Flight with Auto-Pilot System Off 4-66

Task 1114 - Perform Rolling Takeoff 4-67

Task 1155 - Negotiate Wire Obstacles 4-69

Task 1162 - Perform Emergency Egress.. 4-70

Task 1166 - Perform Instrument Maneuvers 4-72

Task 1170 - Perform Instrument Takeoff 4-73

Task 1174 - Perform Holding Procedures.. 4-75

Task 1176 - Perform Nonprecision Approach.. 4-76

Task 1178 - Perform Precision Approach... 4-77

Task 1180 - Perform Emergency Global Positioning System
            Recovery Procedure 4-78

Task 1182 - Perform Unusual Attitude Recovery 4-80

Task 1184 - Respond to Inadvertent Instrument Meteorological Conditions................ 4-81

Task 1188 - Operate Aircraft Survivability Equipment... 4-83

Task 1190 - Perform Hand and Arm Signals 4-85

Task 1194 - Perform Refueling Operations 4-86

Task 1200 - Perform Nonrated Crewmember Duties During
            Maintenance Test Flight.. 4-87

Task 1202 - Perform Auxiliary Power Unit Operations 4-88

Task 1262 - Participate in a Crew-Level After Action Review 4-89

Task 2010 - Perform Multi-Aircraft Operations 4-91

Task 2012 - Perform Tactical Flight Mission Planning... 4-94

Task 2022 - Transmit a Tactical Report... 4-96

Task 2024 - Perform Terrain Flight Navigation.. 4-97

Task 2026 - Perform Terrain Flight 4-99

Task 2036 - Perform Terrain Flight Deceleration... 4-101

Task 2042 - Perform Actions on Contact... 4-102

Task 2048 - Perform External (Sling) Load Operations... 4-105

Task 2052 - Perform Water Bucket Operations 4-108

Task 2060 - Perform Rescue-Hoist/Winch Operations... 4-113

Task 2064 - Perform Paradrop Operations 4-116

Task 2066 - Perform Extended Range Fuel System Operations 4-118

Task 2081 - Operate Night Vision Goggles 4-119

Task 2092 - Respond to Night Vision Goggle Failure 4-120

Task 2112 - Operate Armament Subsystem 4-121

Task 2125 - Perform Pinnacle/Ridgeline Operations 4-123

Task 2127 - Perform Combat Maneuvering Flight 4-125

Task 2169 - Perform Aerial Observation 4-128

Task 4001 - Verify Forms and Records 5-3

Task 4002 - Conduct a Maintenance Test Flight 5-4

Task 4004 - Perform Interior Check 5-5

Task 4010 - Perform Starting Auxiliary Power Unit Check 5-6

Task 4014 - Perform Master Warning Check 5-7

Task 4022 - Perform Brake Check 5-8

Task 4038 - Perform Instrument Display System Check 5-9

Task 4042 - Perform Heater and Vent System Check 5-10

Task 4043 - Perform Windshield Wiper Check 5-11

Task 4044 - Perform Flight Control Hydraulic System Check 5-12

Task 4046 - Perform Flight Collective Friction Check 5-13

Task 4049 - Perform Tail Rotor Pitch Limiter Check 5-14

Task 4064 - Perform Beep Trim Check 5-15

Task 4070 - Perform Fuel Quantity Indicator Check 5-16

Task 4072 - Perform Barometric Altimeter Check 5-17

Task 4073 - Perform Radar Altimeter Check 5-18

Task 4074 - Perform Fire Detection System Check 5-19

Task 4076 - Perform Windshield Anti-Ice Check 5-20

Task 4078 - Perform Pitot Heat Systems Check 5-21

Task 4082 - Perform Fuel Boost Pump Check 5-22

Task 4086 - Perform Engine Starting System Check 5-23

Task 4087 - Perform Engine Abort System Check 5-24

Task 4088 - Perform Starting Engine Check 5-25

Task 4090 - Perform Engine Run-Up System Check 5-26

Task 4091 - Perform Engine Partial Acceleration Check 5-27

Task 4092 - Perform Engine Dust Cover Protector Check 5-28

Task 4093 - Perform Engine Governor Check 5-29

Task 4102 - Perform Electrical System Check 5-30

Task 4112 - Perform Taxi Check 5-31

Task 4119 - Perform Systems Instruments Check 5-32

Task 4142 - Perform Hover Power/Hover Controllability Check...............................5-33

Task 4151 - Perform Auto-Pilot Axis Channel Hold Check..5-34

Task 4193 - Perform In-Flight Check 5-35

Task 4194 - Perform Flight Instruments Check 5-36

Task 4204 - Perform Compasses, Turn Rate, and Vertical Gyros Checks 5-37

Task 4210 - Perform Takeoff and Climb Checks 5-38

Task 4218 - Perform In-Flight Controllability Check 5-39

Task 4226 - Perform Auto-Pilot In-Flight Check 5-40

Task 4236 - Perform Autorotation Revolutions Per Minute Check 5-41

Task 4252 - Perform Vibration Analysis Check 5-42

Task 4254 - Perform Velocity Not to Exceed Check..5-43

Task 4262 - Perform Communication and Navigation Equipment Checks...................5-44

Task 4268 - Perform Cruise Instrument Check 5-45

Task 4274 - Perform In-Flight Communication/Navigation/Flight
            Instruments Check 5-46

Task 4276 - Perform Special Equipment and/or Detailed Procedures Checks.............5-47

Task 4284 - Perform Engine Shutdown Check..5-48

Figures

Figure 4-1. Sample of DA Form 5701-17 4-15
Figure 6-1. Aircrew Coordination Principles 6-2

Tables

Table 2-1. Refresher flight training guide (rated crewmembers) 2-2

Table 2-2. Refresher flight training guide (nonrated crewmember) 2-2

Table 2-3. Mission training task list (rated/nonrated crewmember)...............................2-3

Table 2-4. Task list (rated crewmember) 2-3

Table 2-5. Task list (flight engineer)...2-6

Table 2-6. Task list (nonrated crewmember) 2-9

Table 2–7. Task list (MP/ME minimum evaluation tasks) 2-10

Table 4-1. Sample aircrew briefing checklist 4-6

Table 4-2. Sample nonrated crewmember briefing checklist..4-8

Table 4-3. Sample format for a crew-level after action review checklist........................4-89

Table 4-4. Multi-aircraft operations briefing checklist 4-92

Table 4-5. Sample water bucket guide 4-109

Table 6-1. Examples of standard words and phrases 6-5

Table A-1. Subject area examinations 1

Table A-2. Guide for nonrated crewmember flight training 2

Table A-3. Guide for flight training sequence 3

Table A-4. Guide for flight engineer flight training 3

Preface

This aircrew training manual (ATM) standardizes aircrew training programs (ATPs) and flight evaluation procedures. This manual provides specific guidelines for executing Mi-17 aircrew training. It is based on the training principles outlined at the Army Training Network located on the web at https://atn.army.mil/index.aspx under the Training Management tab. The Mi-17 ATM establishes requirements for crewmember qualification: refresher, mission, and continuation training; and evaluations.

This manual is not a stand-alone document. Requirements of Army regulation (AR) 600-105, AR 600-106, and Training Circular (TC) 3-04.11 must be met. The Kazan Mi-17 flight manual is the authority for operation of the aircraft. If differences exist between the maneuver descriptions in the flight manual and this publication, this publication is the governing authority for training and flight evaluation purposes. Implementation of this manual conforms to AR 95-1 and TC 3-04.11. If a conflict exists between this publication and TC 3-04.11, the ATP commander determines the method of accomplishment based upon the requirement and the unit's mission as to which manual takes precedence.

This manual, in conjunction with AR 600-105, AR 600-106, AR 95-1, and TC 3-04.11, will help develop a comprehensive ATP. Using this ATM ensures that individual crewmember and aircrew proficiency is commensurate with the unit's mission and that aircrews routinely employ standard techniques and procedures.

Crewmembers will use this manual as a "how to" source for performing crewmember duties. It provides performance standards and evaluation guidelines so crewmembers know the level of performance expected. Each task has a description of the proper procedures for completion to meet the standard.

Standardization officers, evaluators, and unit trainers (UTs) will use this manual and TC 3-04.11 as the primary tools in assisting commanders with development and implementation of their ATP.

This publication applies to the Active Army, the Army National Guard (ARNG)/Army National Guard of the United States (ARNGUS), and the United States Army Reserve (USAR), and Department of the Army civilians (DACs) operating the Mi-17 series aircraft, unless otherwise stated.

The proponent for this publication is the United States (U.S.) Army Training and Doctrine Command (TRADOC). Submit comments and recommendations on Department of the Army (DA) Form 2028 (Recommended Changes to Publications and Blank Forms) or automated link via http://www.apd.army.mil through the aviation unit commander to: Commander, U.S. Army Aviation Center of Excellence (USAACE), ATTN: ATZQ-TDT-F, (Flight Training Branch) Building 4507, Joker Street, Fort Rucker, AL 36362-5000, or direct electronic mail questions to: Ruck.ATZQ-TDT-F@conus.army.mil. Recommended changes may also be e-mailed to: ATZQ-ES@rucker.army.mil.

This publication implements portions of Standardization Agreement (STANAG) 3114 (Edition 8).

This publication has been reviewed for operations security considerations.

This page intentionally left blank.

Chapter 1

Introduction

This ATM describes training requirements for crewmembers. It will be used with AR 95-1, AR 600-105, AR 600-106, TC 3-04.11, and other applicable publications. The tasks in this ATM enhance training in individual and aircrew proficiency. This training focuses on tasks supporting the unit's mission. The scope and level of training to be achieved, individually by crewmembers and collectively by aircrews, are dictated by the mission essential task list (METL). Commanders must ensure aircrews are proficient in the METL.

1-1. **CREW STATION DESIGNATION.** The commander will designate a crew station for each crewmember. The individual's commander's task list (CTL) must clearly indicate all crew station designations. Training and proficiency sustainment for rated crewmembers (RCMs) are required in each designated crew station with access to the flight controls. Standardization instructor pilots (SPs), instructor pilots (IPs), instrument examiners (IEs), and aviators designated to fly from both pilot's (PI's) seats will be evaluated, in each seat, during annual proficiency and readiness test (APART) evaluations. Maintenance test pilot evaluators (MEs) and maintenance test pilots (MPs) will follow chapter 5 for crew station requirements and evaluations; however, not all tasks must be evaluated in each seat. Sustainment training for nonrated crewmembers (NCMs) is required in each designated crew station. NCMs are required to be evaluated from all designated crew stations during the APART, but are not required to be evaluated in all tasks from each station. Commanders will develop a program to meet this requirement.

1-2. **SYMBOL USAGE AND WORD DISTINCTIONS.**

    a. **Symbol usage.** The diagonal (/) means one **or** the other **or** both. For example, IP/SP may mean IP **or** SP **or** it may mean IP **and** SP.

    b. **Word distinctions.**

        (1) Warnings, cautions and notes. These words emphasize important and critical instructions.

            (a) *Warning.* A warning is an operating procedure or practice that, if not correctly followed, could result in personal injury or loss of life.

            (b) *Caution.* A caution is an operating procedure or practice that, if not strictly observed, could result in damage to or destruction of equipment.

            (c) *Note.* A note highlights essential information of a non-threatening nature.

        (2) Will, shall, must, should, may, and can. These words distinguish between mandatory, preferred and acceptable methods of accomplishment.

            (a) "Will," "shall," or "must" indicate a mandatory requirement.

            (b) "Should" is used to indicate a non-mandatory but preferred method of accomplishment.

            (c) "May" or "can" is used to indicate an acceptable method of accomplishment.

    c. **Night vision devices (NVDs).**

        (1) Night vision system (NVS) refers to a system attached to the aircraft and is an integral component of the aircraft.

        (2) Night vision goggles (NVG) refers to any image intensifier system; for example, the AN/AVS-6 (aviator's night vision imaging system [ANVIS]).

        (3) NVD refers to NVS and NVG.

    d. **Personnel terminology.**

        (1) The RCM is an aviator; therefore, the terms "rated crewmember," "aviator," and "pilot" are used synonymously.

(2) Pilot in command (PC). The PC has overall responsibility for the operation of the aircraft from pre-mission planning to mission completion and assigns duties to the crew, as necessary. Additionally, the PC is the primary trainer of PIs in the development of experience and judgment.

(3) PI. The PI will complete all tasks assigned by the PC.

(4) UT. The UT is a specialized trainer (RCM or NCM) appointed by the commander to assist with unit training. The UT trains readiness level (RL)-2 crewmembers in mission/additional tasks in accordance with (IAW) the ATM and unit METL. To be qualified as an UT, the crewmember must demonstrate a higher level of knowledge, proficiency, and ability to train other crewmembers IAW the ATM and IPs handbook.

(5) IP. The IP trains and evaluates the RCM and NCM, as appointed by the commander to assist with training. The IP may evaluate an IP/SP during proficiency flight evaluation (PFE) resulting from a lapse in aircraft or NVD currency.

(6) IE. The IE trains and evaluates instrument tasks, as directed by AR 95-1 and local requirements.

(7) SP. The SP trains and evaluates RCM and NCM and supervises and maintains the standardization program.

(8) MP. The MP conducts maintenance test flight (MTF) procedures IAW chapter 5.

(9) ME. The ME trains and evaluates MPs and MEs IAW chapter 5.

(10) NCM. The NCM is a non-aviator who performs operation essential duties aboard an aircraft. NCMs include crew chiefs (CEs), flight engineers (FEs), flight instructors (FIs), and standardization instructors (SIs).

(11) CE. The CE maintains his or her assigned aircraft and performs CE duties.

(12) FE. The FE operates and monitors engine and aircraft systems controls, panels, indicators, and devices. He or she also assists RCMs with operational duties as directed. The commander selects personnel to perform FE duties based on proficiency and experience.

(13) FI. The NCM FI trains and evaluates CEs in aircraft tasks IAW the ATM and unit METL. To qualify as an FI, the crewmember must meet the requirements of AR 95-1.

(14) NCM SI. The SI trains and evaluates CEs, FEs, FIs, and other SIs. The SI assists the unit SP with supervising and maintaining the standardization program. To qualify as an SI, the crewmember must be qualified as an FE or FI and meet the requirements of AR 95-1.

(15) Noncrewmember. These individuals perform duties directly related to the in-flight mission of the aircraft, but not essential to the operation of the aircraft. AR 600-106 lists the categories for noncrewmember positions and the number authorized in each unit. Noncrewmembers may perform CE/FE/UT/FI/SI duties while on noncrewmember flight status if they are military occupational specialty qualified and fully integrated into the commander's ATP. Additionally, noncrewmembers are trained and designated to perform those duties for NCMs who are unable to fly.

Chapter 2

Training

This chapter describes requirements for qualification, RL progression and continuation training. Crewmember qualification requirements will be IAW AR 95-1, TC 3-04.11, and this ATM.

2-1. QUALIFICATION TRAINING.

a. **Initial aircraft qualification.**

(1) RCM. Initial aircraft qualification training in the Mi-17 will be conducted at USAACE or DA-approved training sites IAW a USAACE-approved program of instruction (POI).

(2) NCM. Aircraft qualification training for NCMs is conducted IAW appendix A, applicable regulations, and the commander's ATP.

b. **NVG qualification.** Initial and aircraft NVG qualifications will be conducted IAW TC 3-04.11, the USAACE NVG training support package (TSP), and this ATM. The NVG TSP may be obtained by writing to: Commander, USAACE ATTN: Chief, NVD Branch, 110th Aviation Branch , Fort Rucker, AL 36362-5000 or e-mail: Ruck.ATZQ-ATB-NS@conus.army.mil. The NVG TSP can also be downloaded through Army Knowledge Online from the NVD Branch Knowledge Center at: https://www.us.army.mil/suite/kc/582650.

(1) Initial NVG qualification. Initial qualification will be conducted at the USAACE or DA-approved training site IAW the USAACE-approved POI or locally using the USAACE NVG exportable training package (ETP).

(2) Aircraft NVG qualification.

(a) Academic training. The crewmember will receive training and demonstrate a working knowledge of the appropriate topics outlined in paragraph 3-4b.

(b) Flight training. The crewmember will receive training and demonstrate a working knowledge of the topics in paragraphs 3-4b(7) and (10).

c. Minimum flight hours (RCM). There are no minimum flight hour requirements. This qualification is proficiency based and determined by the crewmember's ability to satisfactorily accomplish the designated tasks.

2-2. REFRESHER TRAINING. Crewmembers are designated RL-3 when they meet the criteria of TC 3-04.11.

a. **Academic training.** The crewmember will receive training and demonstrate a working knowledge of the topics listed in paragraph 3-4b(11).

b. **Flight training.** The crewmember will receive flight training and demonstrate proficiency in all tasks in table 2-4, 2-5 and 2-6 on page 2-3 thru 2-8. Refer to chapter 5 for more guidance.

c. **Minimum flight hours.** There are no minimum flight hour requirements. Training is proficiency based and determined by the crewmember's ability to satisfactorily accomplish the designated tasks. NVG mission training may be included as part of refresher training.

Table 2-1. Refresher flight training guide (rated crewmembers)

Flight Instruction	Hours
Day and night base task training	6.0
Flight evaluation	2.0
*Instrument base task training (aircraft/simulator)	2.0
Instrument evaluation	2.0
Total hours	**12.0**
*Recommend a minimum of 2 hours of instrument base task training be in the aircraft.	

Table 2-2. Refresher flight training guide (nonrated crewmember)

Flight Instruction	Hours
Day and night base task training	6.0
Flight evaluation	2.0
Total hours	**8.0**

    d. **NVG refresher training**.

(1) Academic training. The crewmember will receive training and demonstrate a working knowledge of the applicable topics in paragraphs 3-4b(7) and (10).

(2) Flight training. The crewmember will receive training and demonstrate proficiency in all base tasks marked with an "**X**" in the NVG column of table 2-4, page 2-3, or table 2-5, page 2-6, as applicable. The commander may select additional base tasks.

(3) Minimum flight hours. There are no minimum flight hour requirements. The training is proficiency based and determined by the crewmember's ability to satisfactorily accomplish the designated tasks.

2-3. **MISSION TRAINING**. Crewmembers are designated RL 2 when they meet the criteria of TC 3-04.11.

    a. Training requirements.

(1) Mission training. Mission training programs help RL-2 crewmembers develop the ability to perform specific tasks selected by the commander to support the unit's METL.

(a) Academic training. The crewmember will receive training and demonstrate a working knowledge of the topics listed in paragraphs 3-4b(8) and (9).

(b) Flight training. The training will consist of those mission tasks in table 2-3, page 2-3, as selected by the commander and additional tasks necessary to complete the unit's mission. This training may be conducted by a UT. The crewmember will receive training from all designated crew stations. A task performed from either crew station does not need to be evaluated from both stations. Flight mission-training hour requirements are based on demonstrated proficiency. The evaluation must be conducted by an SP, IP, SI, or FI and may be continuous.

    b. The bolded tasks found in the following tables represent the performance tasks for Mi-17 aircraft.

Table 2-3. Mission training task list (rated/nonrated crewmember)

Task	Task Title
2010	Perform Multi-Aircraft Operations
2022	Transmit a Tactical Report
2024	**Perform Terrain Flight Navigation**
2026	**Perform Terrain Flight**
2036	**Perform Terrain Flight Deceleration**
2042	Perform Actions on Contact
2048	**Perform External (Sling) Load Operations**
2052	**Perform Water Bucket Operations**
2060	**Perform Rescue Hoist Operations**
2064	**Perform Para-Drop Operations**
2066	**Perform Extended Range Fuel System Operations**
2081	**Operate Night Vision System Operations**
2092	**Respond to Night Vision Google Failure**

Table 2-4. Task list (rated crewmember)

Legend
D – Tasks performed during day flight.
I – Tasks performed during instrument flight.
N – Tasks performed during unaided night flight. The tasks selected in this column do not need to be evaluated during the standardization evaluation. If tasks are evaluated at night, it will suffice for tasks required in day conditions.
NVG – Tasks performed during NVG flight. Tasks evaluated while using NVG will suffice for tasks required in day conditions.
S, I, or NG in the EVAL column – Mandatory tasks for standardization, instrument, or annual NVG flight evaluations respectively.

Task	Task Title	D	I	N	NVG	EVAL
1000	Participate in a Crew Mission Briefing			X		S,I,NG
1002	Conduct Passenger Briefing			X		S or NG
1004	Plan a Visual Flight Rules Flight	X		X	X	S
1006	Plan an Instrument Flight Rules Flight		X			I
1010	Prepare a Performance Planning Card			X		S
1012	Verify Aircraft Weight and Balance			X		S
1013	**Operate Mission Planning System**			X		S, NG
1016	Perform Internal Load Operations			X		S
1022	Perform Preflight Inspection			X		S
1024	**Perform Before-Starting Engine Through Before-Leaving Helicopter Checks**			X		S, NG
1026	**Maintain Airspace Surveillance**			X		S, NG
1028	**Perform Hover/Power Check**			X		S,I,NG
1032	Perform Radio Communication Procedures			X		S, I
1034	**Perform Ground Taxi**	X		X	X	S, NG
1038	**Perform Hovering Flight**	X		X	X	S,NG

Table 2-4. Task list (rated crewmember) (cont.)

Legend
D – Tasks performed during day flight.
I – Tasks performed during instrument flight.
N – Tasks performed during unaided night flight. The tasks selected in this column do not need to be evaluated during the standardization evaluation. If tasks are evaluated at night, it will suffice for tasks required in day conditions.
NVG – Tasks performed during NVG flight. Tasks evaluated while using NVG will suffice for tasks required in day conditions.
S, I, or NG in the EVAL column – Mandatory tasks for standardization, instrument, or annual NVG flight evaluations respectively.

Task	Task Title	D	I	N	NVG	EVAL
1040	**Perform Visual Meteorological Conditions Takeoff**	X		X	X	S, NG
1044	Navigate by Pilotage and Dead Reckoning	X		X	X	S, NG
1046	Perform Electronically Aided Navigation			X		S
1048	Perform Fuel Management Procedures			X		S,I,NG
1052	**Perform Visual Meteorological Conditions Flight Maneuvers**	X		X	X	S, NG
1054	Select Landing Zone/Pickup Zone/Holding Area	X		X	X	S, NG
1058	**Perform Visual Meteorological Conditions Approach**	X		X	X	S, NG
1062	**Perform Slope Operations**	X			X	S, NG
1064	**Perform Roll-On Landing**	X		X	X	S, NG
1068	**Perform Go-Around**	X		X	X	S, NG
1070	**Respond To Emergencies**	X	X	X	X	S, NG
1074	**Respond to Engine Failure at Cruise Flight**	X		X	X	S, NG
1075	**Perform Single-Engine Landing**	X			X	S,NG
1082	**Perform Autorotation**	X		X	X	S, NG
1094	**Perform Flight with Auto-Pilot System Off**	X			X	S, NG
1114	**Perform a Rolling Takeoff**	X		X	X	S, NG
1155	Negotiate Wire Obstacles	X			X	S, NG
1162	Perform Emergency Egress	X				S
1166	Perform Instrument Maneuvers		X			I
1170	**Perform Instrument Takeoff**		X			I
1174	Perform Holding Procedures		X			I
1176	**Perform Nonprecision Approach**		X			I
1178	**Perform Precisions Approach**		X			I
1180	Perform Emergency Global Positioning System Recovery Procedure		X			
1182	**Perform Unusual Attitude Recovery**	X	X		X	S or I
1184	**Respond to Inadvertent Instrument Meteorological Conditions**	X		X	X	S, NG
1188	Operate Aircraft Survivability Equipment	X				S
1190	Perform Hand and Arm Signals	X				
1194	Perform Refueling Procedures			X		
1202	Perform Auxiliary Power Unit Operations			X		S

Table 2-4. Task list (rated crewmember) (cont.)

Legend

D – Tasks performed during day flight.

I – Tasks performed during instrument flight.

N – Tasks performed during unaided night flight. The tasks selected in this column do not need to be evaluated during the standardization evaluation. If tasks are evaluated at night, it will suffice for tasks required in day conditions.

NVG – Tasks performed during NVG flight. Tasks evaluated while using NVG will suffice for tasks required in day conditions.

S, I, or NG in the **EVAL** column – Mandatory tasks for standardization, instrument, or annual NVG flight evaluations respectively.

Task	Task Title	D	I	N	NVG	EVAL
1262	Participate in a Crew-Level After Action Review			X		S, I, NG
2081	Operate Night Vision Goggles				X	NG
2092	**Respond to Night Vision Goggle Failure**				X	NG

(2) NVG mission training. NVG mission training will be IAW the commander's training program which specifies tasks. When commanders determine a requirement for using NVG in mission profiles, they must specify mission tasks to support the unit's METL. Before undergoing NVG mission training, the crewmember must complete qualification or refresher training and must be NVG current.

(a) Academic training. The crewmember will receive training and demonstrate a working knowledge of the subject areas in paragraphs 3-4b(7) through (10) and additional subject areas selected by the commander.

(b) Flight training. The crewmember will receive flight training and demonstrate proficiency in the mission and additional NVG tasks, as specified on the task list for the crewmember's position.

(3) MP/ME mission training. MPs and MEs should be limited to duties in one primary and one alternate (or additional) aircraft. The MP/ME will complete tasks outlined in table 2-6, page 2-8, and should be required to complete those mission/additional tasks selected by the commander. Crewmembers undergoing training in the aircraft must fly with an ME for maintenance training.

(a) Academic training. The MP will receive training and demonstrate a working knowledge of the topics listed in paragraph 3-4b(11).

(b) Flight training. The MP/ME will receive flight training and demonstrate proficiency in the tasks found in table 2-6, page 2-8. Refer to chapter 5 for more guidance.

c. Minimum flight hours. There are no minimum flight hour requirements. Training is proficiency based and determined by the crewmember's ability to satisfactorily accomplish the designated tasks. NVG mission training may be included as part of refresher training.

2-4. **CONTINUATION TRAINING.** Crewmembers are designated RL 1 when they meet the criteria of TC 3-04.11.

Note. UTs and evaluators may credit those hours they fly while performing assigned duties, regardless of their crew station, toward their semi-annual flying-hour requirements.

---

a. Semi-annual flying-hour requirements–Aircraft. The minimum requirements for crewmembers are as follows:

(1) RCMs.

(a) Flight activity category (FAC) 1–45 hours, which must be flown while occupying a crew station with access to the flight controls.

(b) FAC 2–35 hours, which must be flown while occupying a crew station with access to the flight controls.

(c) FAC 3–No flying-hour requirements.

(2) NCMs–24 hours, in the aircraft while performing crew duties.

b. Semiannual flying-hour requirements–NVG. The commander will determine semiannual flying-hour requirements for NVG. The requirement will be tailored to the individual crewmember based on proficiency and experience. RCMs will complete the requirements in the aircraft while occupying a crew station with access to the flight controls. NCMs will complete the requirements while performing crew duties.

c. Annual flight simulator (FS) device flying-hour requirements. All aviators and FEs within 200 statute miles (SMs) of a compatible synthetic flight training system (SFTS) device will complete the following number of hours in the SFTS. The commander will determine FS requirements for RCMs outside of 200 SMs. RCMs/FEs may apply 12 hours of Mi-17 FS time toward their semiannual flying-hour requirement. Time flown in non-compatible FSs will not be credited towards the minimum annual flying hour or FS requirements (AR 95-1, paragraph 4-11d). The only compatible FS is the Mi-17 mission-designated symbol.

(1) FAC 1–18 hours annually.

(2) FAC 2 and FE–12 hours annually.

(3) FAC 3–10 hours semiannually regardless of distance from the FS.

d. Annual task and iteration requirements. The minimum requirements are as follows:

(1) FAC 1 and FAC 2. Each crewmember must perform at least one task iteration annually in each required flying mode as indicated in table 2-4, page 2-3, or table 2-5, page 2-6, the tasks selected from table 2-3, page 2-3, and additional tasks on the CTL. One iteration of each task must be performed in the aircraft. Tasks performed at night (or while using NVG) may be counted for day iterations. The crewmember is responsible for maintaining proficiency in each task. The commander may require additional iterations of specific tasks.

Table 2-5. Task list (flight engineer)

Legend
D – Tasks performed during day flight.
I – Tasks performed during instrument flight.
N – Tasks performed during unaided night flight. The tasks selected in this column do not need to be evaluated during the standardization evaluation. If tasks are evaluated at night, it will suffice for tasks required in day conditions.
NVG – Tasks performed during NVG flight. Tasks evaluated while using NVG will suffice for tasks required in day conditions.
S, I, or **NG** in the **EVAL** column – Mandatory tasks for standardization, instrument, or annual NVG flight evaluations respectively.

Task	Task Title	D	I	N	NVG	EVAL
1000	Participate in a Crew Mission Briefing			X		S,I,NG
1002	Conduct Passenger Briefing			X		S or NG
1004	Plan a Visual Flight Rules Flight	X		X	X	S
1010	Prepare a Performance Planning Card			X		S
1012	Verify Aircraft Weight and Balance			X		S
1016	Perform Internal Load Operations			X		S
1022	Perform Pre-Flight Inspection			X		S or I
1024	**Perform Before Starting Engine through Before Leaving Helicopter Checks**			X		S, NG
1026	**Maintain Airspace Surveillance**			X		S, NG
1028	Perform Hover/Power Check			X		S,I,NG
1032	Perform Radio Communication Procedures			X		S, I
1034	**Perform Ground Taxi**	X		X	X	S, NG
1038	**Perform Hover Flight**	X		X	X	S,NG

Table 2-5. Task list (flight engineer) (cont.)

Legend
D – Tasks performed during day flight.
I – Tasks performed during instrument flight.
N – Tasks performed during unaided night flight. The tasks selected in this column do not need to be evaluated during the standardization evaluation. If tasks are evaluated at night, it will suffice for tasks required in day conditions.
NVG – Tasks performed during NVG flight. Tasks evaluated while using NVG will suffice for tasks required in day conditions.
S, I, or NG in the **EVAL** column – Mandatory tasks for standardization, instrument, or annual NVG flight evaluations respectively.

Task	Task Title	D	I	N	NVG	EVAL
1040	**Perform Visual Meteorological Conditions Takeoff**	X		X	X	S, NG
1044	Navigate by Pilotage and Dead Reckoning	X		X	X	S, NG
1046	Perform Electronically Aided Navigation			X		S
1048	Perform Fuel Management Procedures			X		S,I,NG
1052	Perform Visual Meteorological Conditions Flight Maneuvers	X		X	X	S, NG
1054	Select Landing Zone/Pickup Zone/Holding Area	X		X	X	S, NG
1058	**Perform Visual Meteorological Conditions Approach**	X		X	X	S, NG
1062	**Perform Slope Operations**	X			X	S, NG
1064	**Perform Roll-On Landing**	X		X	X	S, NG
1068	Perform Go-Around	X		X	X	S, NG
1070	**Respond to Emergencies**	X	X	X	X	S, NG
1074	**Respond to Engine Failure at Cruise Flight**	X		X	X	S, NG
1082	Perform Autorotation	X		X	X	S, NG
1094	Perform Flight with Auto-Pilot System Off	X			X	S,
1114	Perform Rolling Takeoff	X		X	X	S, NG
1155	Negotiate Wire Obstacles	X			X	S, NG
1162	Perform Emergency Egress	X				S
1166	Perform Instrument Maneuvers		X			I
1170	Perform Instrument Takeoff		X			I
1174	Perform Holding Procedures		X			I
1176	Perform Nonprecision Approach		X			I
1178	Perform Precision Approach		X			I
1180	Perform Emergency Global Positioning System Recovery Procedure		X			
1182	Perform Unusual Attitude Recovery	X	X		X	S, I
1184	Respond to Inadvertent Instrument Meteorological Conditions	X		X	X	S, NG
1188	Operate Aircraft Survivability Equipment	X				S
1194	Perform Refueling Procedures			X		
1202	Perform Auxiliary Power Unit Operations			X		S
1262	Participate in a Crew-Level After Action Review	X	X	X	X	S,I,NG
2081	Operate Night Vision Goggles				X	NG

Table 2-5. Task list (flight engineer) (cont.)

Legend
D – Tasks performed during day flight.
I – Tasks performed during instrument flight.
N – Tasks performed during unaided night flight. The tasks selected in this column do not need to be evaluated during the standardization evaluation. If tasks are evaluated at night, it will suffice for tasks required in day conditions.
NVG – Tasks performed during NVG flight. Tasks evaluated while using NVG will suffice for tasks required in day conditions.
S, **I**, or **NG** in the **EVAL** column – Mandatory tasks for standardization, instrument, or annual NVG flight evaluations respectively.

Task	Task Title	D	I	N	NVG	EVAL
2092	Respond to Night Vision Goggle Failure				X	NG

Table 2-6. Task list (nonrated crewmember)

Legend

D – Tasks performed during day flight.

N – Tasks performed during unaided night flight. The tasks selected in this column do not need to be evaluated during the standardization evaluation. If tasks are evaluated at night, it will suffice for tasks required in day conditions.

NVG – Tasks performed during NVG flight. Tasks evaluated while using NVG will suffice for tasks required in day conditions.

S or NG in the **EVAL** column – Mandatory tasks for standardization, instrument, or annual NVG flight evaluations respectively.

Task	Task Title	D	N	NVG	EVAL
1000	Participate in a Crew Mission Briefing		X		S
1012	Verify Aircraft Weight and Balance		X		
1014	Operate Aviation Life Support Equipment		X		S
1016	Perform Internal Load Operations		X		S
1019	Perform Preventive Maintenance Daily Check		X		S
1022	Perform Pre-Flight Inspection		X		S
1024	**Perform Before Starting Engine through Before Leaving Helicopter Checks**		X		S
1026	**Maintain Airspace Surveillance**		X		S, NG
1028	Perform Hover/Power Check		X		S
1032	Perform Radio Communications Procedures		X		S
1034	**Perform Ground Taxi**		X		S
1038	**Perform Hovering Flight**	X	X	X	S, NG
1040	**Perform Visual Meteorological Conditions Takeoff**		X		S
1048	Perform Fuel Management Procedures		X		S
1058	**Perform Visual Meteorological Conditions Approach**	X	X	X	S, NG
1062	**Perform Slope Operations**	X	X	X	S, NG
1064	**Perform Roll-On Landing**	X	X	X	S, NG
1070	**Respond to Emergencies**	X	X	X	S, NG
1190	Perform Hand and Arm Signals		X		S
1194	Perform Refueling Operations		X		S
1200	Perform NCM Duties During Maintenance Test Flight	X			
1202	Perform Auxiliary Power Unit Operations		X		S
1262	Participate in a Crew-Level After Action Review	X	X	X	S
2081	Operate Night Vision Goggles			X	NG
2092	**Respond to Night Vision Goggle Failure**			X	NG

(2) FAC 3. Each crewmember must perform, annually, at least one iteration of each task annotated on the CTL in the FS. The crewmember is responsible for maintaining proficiency in each task. The commander may require additional iterations of specific tasks.

(3) MPs and MEs. In addition to the required minimum annual tasks and iterations, MPs and MEs will perform a minimum of four iterations of MTF tasks annually (table 2-7, page 2-9). MEs will perform a minimum of two of the four iterations mentioned above from each flight crew station with access to the flight controls.

Table 2–7. Task list (MP/ME minimum evaluation tasks)

Task	Task Title
4088	Perform Starting Engine Check
4090	Perform Engine Run-Up System Check
4112	Perform Taxi Check
4142	Perform Hover Power/Hover Controllability Check
4193	Perform In-Flight Check
4204	Perform Compasses, Turn Rate and Vertical Gyros Check
4236	Perform Autorotation Revolutions Per Minute Check
4274	Perform In-Flight Communication/Navigation/Flight Instruments Check
4276	Perform Special Equipment and/or Detailed Procedures Check
4284	Perform Engine Shutdown Checks

e.   Hood/weather requirements. All aviators will complete hood or weather requirements as determined by the commander. This requirement may be completed in the aircraft or FS.

2-5. **TASK LIST.**

a.   Performance tasks. For the purpose of clarifying mode and conditions, a performance task is differentiated from a technical task. An ATM performance task is significantly affected by the conditions and the mode of flight. The mode and condition under which the task must be performed is specified.

b.   Technical tasks. Technical tasks measure a crewmember's ability to plan an action such as a flight, pre-flight, participate in crew mission briefing, and perform hover power check. Technical tasks are not significantly affected by the mode of flight and may be performed or evaluated in any mode.

Note. The requirement to perform instrument tasks in additional aircraft, in category, will be at the discretion of the commander.

Note. RCMs required to perform MP or ME duties in the Mi-17 as an additional or alternate aircraft will perform four iterations of the required tasks.

c.   Base tasks. Table 2-4, page 2-4; table 2-5, page 2-6; and table 2-6, page 2-8 list RCM, FE, and NCM base task requirements.

d.   Mission tasks. Table 2-3, page 2-3, lists RCM and NCM mission tasks. The commander will select mission and additional tasks and iterations supporting the unit's METL and individual proficiency. The commander will determine evaluation requirements for all mission tasks and modes of flight and annotate the aircrew member's CTL accordingly.

e.   MP tasks. Refer to chapter 5.

f.   Evaluation guidelines. Aviators designated to fly from both pilot seats are evaluated, in each seat, during APART evaluations. However, not all tasks must be evaluated from each crew station. Sustainment training for NCMs is required in each designated crew station. CEs/FIs/SIs are required to be evaluated from the cabin door position in the aircraft during the APART, but are not required to be evaluated in all tasks from each position. FEs/SIs is required to be evaluated from the jump seat, during APART evaluations. Other positions may be evaluated at the discretion of the evaluator. APART and annual evaluation tasks are designated by an S, I, and/or NG in the EVAL column of tables 2-4 and 2-5. The tasks selected under the "N" column do not need to be evaluated during the standardization evaluation. Tasks evaluated at night (or while using NVG) will suffice for tasks required in day conditions. Mission tasks will be evaluated during the APART if the task is on the individual's CTL and designated with an "E" for evaluation. The commander should select mission/additional mission tasks for evaluation based on the unit's METL. Refer to chapter 5 for MP/ME APART requirements.

2-6. CURRENCY REQUIREMENTS.

a. Aircraft currency. Aircraft currency will be IAW AR 95-1. FE/SI will follow RCM requirements IAW AR 95-1. A crewmember with lapsed currency must complete a PFE, administered by an evaluator in the aircraft. The crewmember will demonstrate proficiency in those tasks and modes selected by the commander. If the crewmember fails to demonstrate proficiency, he or she will be placed in the appropriate RL. An appropriate training program will be developed to enable the crewmember to regain proficiency in the unsatisfactory tasks.

b. NVG currency. To be considered NVG current, crewmembers will participate, at least once every 60 consecutive days, in a 1-hour flight in the aircraft while wearing NVG. RCMs will occupy a crew station with access to the flight controls. NCMs must be performing crew duties.

(1) Crewmember. If a crewmember's currency has lapsed, the crewmember must complete (at a minimum) a 1-hour NVG PFE administered at night in the aircraft by an NVG SP, IP, SI, or FI, as appropriate.

(2) RCM. The RCM must occupy a crew station with access to the flight controls during the evaluation.

(3) NCM. The NCM must occupy a crew station in the aircraft while performing crew duties during the evaluation.

(4) Minimum tasks. Minimum tasks to be evaluated are indicated by an "NG" in the EVAL column of table 2-4, page 2-4, or table 2-5, page 2-6 and 2-7 as applicable. The commander may designate other mission and/or additional tasks.

2-7. CHEMICAL, BIOLOGICAL, RADIOLOGICAL, NUCLEAR, AND HIGH YIELD EXPLOSIVE TRAINING. IAW TC 3-04.11, crewmembers must wear the complete chemical, biological, radiological, nuclear, and high yield explosive (CBRNE) ensemble during CBRNE training. All CBRNE training will be performed in the aircraft. CBRNE training is not required for FAC 3 positions and DACs.

CAUTION

While conducting CBRNE training, the commander will ensure that aircrews exercise caution while performing flight duties when the wet bulb globe temperature is above 75 degrees Fahrenheit.

---

a. RCM tasks. RCMs will receive CBRNE training in the following tasks. The commander may select other tasks based on the unit mission.

(1) Task 1024, Perform Before-Starting Engine Through Before-Leaving Helicopter Check.

(2) Task 1028, Perform Hover Check.

(3) Task 1040, Perform Visual Meteorological Conditions Takeoff.

(4) Task 1058, Perform Visual Meteorological Conditions Approach.

(5) Task 2036, Perform Terrain Flight Deceleration.

b. NCM tasks. NCMs will receive CBRNE training in Task 1024. The commander may select other tasks based on the unit mission.

This page intentionally left blank.

Chapter 3

Evaluation

This chapter describes evaluation principles and grading considerations for individual crewmembers. It also contains guidelines for conducting academic and hands-on performance testing. Evaluations are a primary means of assessing flight standardization and crewmember proficiency. Evaluations will be conducted IAW AR 95-1, the commander's ATP, TC 3-04.11, and this ATM.

3-1. EVALUATION PRINCIPLES. The value of any evaluation depends on adherence to fundamental evaluation principles. These principles are described below.

    a. **Selection of evaluators.** The evaluators must be selected not only for their technical qualifications but also for their demonstrated performance, objectivity, and ability to observe and to provide constructive comments. These evaluators are the SPs, IPs, IEs, MEs, SIs, and FIs who assist the commander in administering the ATP.

    b. **Method of evaluation.** The method used to conduct the evaluation must be based on uniform and standard objectives. In addition, it must be consistent with the unit's mission and must strictly adhere to the appropriate standing operating procedures (SOPs) and regulations. The evaluator must ensure a complete evaluation is given in all areas and refrain from making a personal "area of expertise" a dominant topic during the evaluation.

    c. **Participant understanding.** All participants must completely understand the purpose of the evaluation.

    d. **Participant cooperation.** All participants must cooperate to guarantee the accomplishment of the evaluation objectives. The emphasis is on all participants, not just on the examinee.

    e. **Identification of training needs.** The evaluation must produce specific findings to identify training needs. Any crewmember affected by the evaluation needs to know what is being performed correctly and incorrectly and how improvements can be made.

    f. **Purpose of evaluation.** The evaluation determines the examinee's ability to perform essential hands-on/academic tasks to prescribed standards. The flight evaluation will also determine the examinee's ability to exercise crew coordination in completing these tasks.

    g. **Aircrew coordination.** The guidelines for evaluating aircrew coordination are based on a subjective analysis of how effectively a crew performs collectively to accomplish a series of tasks. The evaluator must determine how effectively the examinee employs aircrew coordination, as outlined in chapter 6 of this ATM.

    h. **Evaluator role as crewmember.** In all phases of evaluation, the evaluator is expected to perform as an effective crewmember. However, at some point during the evaluation, circumstances may prevent the evaluator from performing as an effective crewmember.

    (1) In such cases, a realistic, meaningful, and planned method should be developed to effectively pass this task back to the examinee. In all other situations, the evaluator must perform as outlined in the task description or as directed by the examinee to determine the examinee's level of proficiency; the evaluator may intentionally perform as an ineffective crewmember.

    (2) During the flight evaluation, the evaluator will normally perform as outlined in the task description or as directed by the examinee. At some point, the evaluator may perform a role reversal with the examinee. The examinee must be informed of the initiation and termination of role reversals. The examinee must know when they are supported by a fully functioning crewmember.

Note. When evaluating a PC, SP, IP, ME, UT, or IE, the evaluator must advise the examinee that, during role-reversal, the evaluator may deliberately perform some tasks or aircrew coordination outside the standards to check the examinee's diagnostic and corrective action skills.

3-2. **GRADING CONSIDERATIONS.**

    a. **Academic evaluation.** The examinee must demonstrate a working knowledge and understanding of the appropriate subject areas in paragraph 3-4b.

    b. **Flight evaluation.**

        (1) Academic. Some tasks are identified as tasks that may be evaluated academically. The examinee must demonstrate a working knowledge of these task. Evaluators may use computer-based instruction, mock-ups, or other approved devices (to include the aircraft or FS) to determine the examinee's knowledge of the task.

        (2) In the aircraft or the FS. Those tasks requiring evaluation under these conditions must be perform in the aircraft or the Mi-17 FS. Task standards are based on an ideal situation; grading is based on meeting the minimum standards. The evaluator must consider deviations (high wind, turbulence, or poor visibility) from the ideal during the evaluation. If other than ideal conditions exist, the evaluator must make appropriate adjustments to the standards.

Note. During an evaluation, a task iteration performed in a more demanding mode of flight may suffice for an iteration performed in a less demanding mode of flight. The commander determines which mode of flight is more demanding.

3-3. **CREWMEMBER EVALUATION.** Evaluations are conducted to determine the crewmember's ability to perform the tasks on the CTL and check understanding of required academic subjects listed in the ATM. The evaluator will determine the amount of time devoted to each phase. When the examinee is an evaluator/trainer or a UT, the recommended procedure is for the evaluator to reverse roles with the examinee. When the evaluator uses this technique, the examinee must understand how the role-reversal will be conducted and when it will be in effect.

Note. Initial validation of a crewmember's qualifications following an additional skill identifier producing course of flight instruction/school (such as the Mi-17 IP, MP, IE, or FE course) will be conducted in the aircraft.

---

    a. **Recommended performance and evaluation criteria.**

        (1) PI. The PI must demonstrate a working knowledge of the appropriate subjects in paragraph 3-4b. In addition, the PI must be familiar with the individual aircrew training folder (IATF) and understand the requirements of the CTL.

        (2) PC/MP. The PC/MP must meet the requirements in a paragraph 3-3a (1). In addition, the PC/MP must demonstrate sound judgment and maturity in the management of the mission, crew, and assets.

        (3) UT. The UT must meet the requirements in paragraph 3-3a (2) or (8). In addition, the UT must be able to instruct in the appropriate tasks and subjects, recognize errors in performance or understanding, make recommendations for improvement, train to standards, and document training.

        (4) IP or IE. The IP or IE must meet the requirements in paragraph 3-3a (2). In addition, the IP/IE must be able to objectively train, evaluate, and document performance of the UT, PC, PI, SI, FI, FE and CE using role-reversal as appropriate. This individual must possess a thorough knowledge of the fundamentals of instruction and evaluation, be able to develop and implement an individual training plan, and possess a thorough understanding of the requirements and administration of the ATP.

        (5) SP/IE. The SP/IE must meet the requirements in paragraph 3-3a (2) and (4). The SP/IE must be able to train and evaluate SPs, IPs, IEs, UTs, PCs, PIs, SIs and FIs using role reversal as appropriate. The SP must also be able to develop and implement a unit-training plan and administer the commander's ATP.

        (6) ME. The ME must meet the requirements in paragraph 3-3a(2). The ME must be able to train and evaluate other MEs and functional check pilots using role reversal as appropriate. The ME must possess a thorough knowledge of the fundamentals of instruction and evaluation.

        (7) CE. The CE must demonstrate an understanding of conditions, standards, descriptions, and appropriate considerations on the CTL. The CE must perform selected tasks to ATM standards while applying aircrew coordination. The CE must also demonstrate a basic understanding of the appropriate academic subjects listed in paragraph 3-4b, be familiar with the IATF, and understand the requirements of the CTL.

(8) FE. The FE must perform selected tasks to ATM standards while applying aircrew coordination. The FE must demonstrate sound judgment, and technical/tactical proficiency in the employment of the aircraft, the unit's mission, crew, and assets.

(9) FI. The FI must meet the requirements in paragraph 3-3a(7). In addition, the FI must be able to objectively train, evaluate, and document the performance of the UTs, CEs, and observers (ORs) (aircraft maintenance personnel, technical OR, gunner, or other personnel performing duties requiring flight) as appropriate; be able to develop and implement an individual training plan; and have a thorough understanding of the requirements and administration of the ATP.

(10) *SI. The SI must meet the requirements in paragraphs 3-3a(7 and 9). In addition, the SI must be able to train and evaluate SIs, FIs, UTs, FEs, CEs, and ORs as appropriate; and be able to develop and implement a unit training plan; and administer the commander's ATP for NCMs.

Note. In order for a SI to evaluate an FE, the SI must be a current and qualified FE.

Note. Evaluators/trainers will be evaluated on their ability to apply the fundamentals of instruction as outlined in paragraph 3-4b(12).

Note. During academic evaluations, evaluators should ask questions addressing specific topics in each area and avoid those requiring laundry list-type answers. Questions should be developed as described in the IP's handbook.

b. **Academic evaluation criteria.**

(1) PFE. The SP/IP/SI/FI will evaluate appropriate subject areas in paragraph 3-4b.

(2) APART standardization/annual NVG evaluations. The SP/IP/SI/FI will evaluate a minimum of two topics from each applicable subject area in paragraph 3-4b.

(3) APART instrument evaluation. The IE will evaluate a minimum of two topics from the subject areas in paragraphs 3-4b(1) through 3-4b(5), relative to instrument flight rules (IFR) and flight planning. If the evaluated crewmember is an IP/SP/IE, the IE will evaluate the ability of the IP/SP/IE to instruct instrument-related areas or subjects.

(4) APART MP/ME evaluation. The ME will evaluate a minimum of two topics from the applicable subject areas in paragraph 3-4b, emphasizing how they apply to MTFs.

(5) Other ATP evaluations. The SP/IP/SI/FI will evaluate appropriate subject areas in paragraph 3-4b.

3-4. **EVALUATION SEQUENCE.** The evaluation sequence consists of four phases–Introduction, Academic Evaluation Topics, Flight Evaluation, and Debriefing. The evaluator will determine the amount of time devoted to each phase.

a. **Phase 1–Introduction.** In this phase, the evaluator—

(1) Reviews the examinee's individual flight record folder and IATF record to verify that the examinee meets all prerequisites for designation and has a current DA Form 4186 (Medical Recommendation for Flying Duty)

(2) Confirms the purpose of the evaluation, explains the evaluation procedure, and discusses evaluation standards and criteria to be used.

b. **Phase 2–Academic Evaluation Topics.**

(1) Regulations and publications (AR 95-1, AR 95-2, Federal Aviation regulations [FARs], Department of the Army Pamphlet [DA Pam] 738-751, Department of Defense flight information publications [DOD FLIPs]), TC 3-04.11, technical manual [TM] 1-1500-328-23, Program Management-Nonstandard Rotary-Wing Aircraft [PM-NSRWA] approved flight and maintenance manuals, and local and unit SOPs). Topics in this subject area are—

- ATP requirements.
- Crew coordination.
- Airspace regulations and usage.
- Flight plan preparation and filing.

- Performance planning.
- Inadvertent instrument meteorological conditions (IIMC) procedures.
- Forms, records, and publications required in the aircraft.
- Unit SOP and local requirements.
- DOD FLIPs and maps.
- Visual flight rules (VFR)/IFR minimums and procedures.
- Weight and balance requirements.
- Maintenance forms and records.
- Aviation life support equipment (ALSE).

(2) Aircraft systems, avionics, mission equipment description and operation. Topics in this subject area are—

- Engines and related systems.
- Emergency equipment.
- Transponder.
- Fuel system.
- Power train system.
- Flight control.
- Hydraulic/pneumatic systems.
- Utility system.
- Rotor system.
- Flight instruments.
- Airframe and landing gear system.
- Auxiliary power unit (APU).
- Lighting.
- Aircraft survivability equipment (ASE).
- Servicing, parking and mooring.
- Cargo handling systems.
- Mission equipment.
- Armament.
- Avionics.
- Auto-pilot system.
- Heating, ventilation, cooling, and environmental control unit.
- Electrical power supply and distribution system.

(3) Operating limitations and restrictions. Topics in this subject area are—

- Wind limitations.
- Rotor limitations.
- Power limitations.
- Engine limitations.
- Aircraft system limitations.
- Airspeed limitations.
- Temperature limitations.
- Loading limitations.
- Weapon system limitations.
- Maneuvering limits.
- Flight envelope limitations (such as extended range fuel system, cargo/rescue hoist, external/internal load operations).

- Weather requirements.
- Environmental limitations/restrictions.

(4) Aircraft emergency procedures and malfunction analysis. Topics in this subject area are—

- Emergency terms and their definitions.
- Engine malfunctions.
- Fires.
- Hydraulic/pneumatic system malfunctions.
- Landing and ditching procedures.
- Mission equipment malfunctions.
- Rotor, transmission, and drive train system malfunctions.
- Emergency exits and equipment.
- Chip detectors.
- Fuel system malfunctions.
- Electrical system malfunctions.
- Flight control malfunctions.
- Auto-pilot malfunctions.

(5) Aeromedical factors (AR 40-8, , FM 3-04.203, and TC 3-04.93-. Topics in this subject area are—

- Flight restrictions due to exogenous factors.
- Stress and fatigue.
- Spatial disorientation.
- Altitude psychology.
- Hypoxia.
- Middle ear discomfort.
- Principles and problems of vision.

(6) Aerodynamics (FM 3-04.203). This subject area applies only to RCMs. Topics in this subject area are—

- Attitude and heading control.
- Dissymmetry of lift.
- In-ground effect/out-of-ground effect (OGE) hovering flight.
- Characteristics of dynamic roll over.
- Loss of tail rotor effectiveness.
- Retreating blade stall.
- Effective translational lift.
- Settling with power.
- Types of drag.

(7) Night mission operations (FM 3-04.203 and TC 3-04.93). Topics in this subject area are—

- Unaided night flight.
- Visual illusions.
- Distance estimation and depth perception.
- Night vision limitations and techniques.
- Types of vision.
- Use of internal and external lights.

(8) Tactical and mission operations (ATTP 3-18.12, FM 3-04.126, FM 3-04.203, FM 3-11, FM 3-52, FM 4-20.197, FM 4-20.198, FM 4-20.199, FM 55-450-2, the commander's ATP, and unit SOP). Topics in this subject area are—

- CBRNE operations.
- ASE employment.

- Downed aircraft procedures.
- Aircraft armament subsystems.
- Communication security (COMSEC).
- Mission equipment.
- Internal load operations.
- Aviation mission planning.
- Fratricide prevention.
- Evasive maneuvers.
- Cargo/rescue hoist operations.
- External (sling) load operations.
- High-intensity radio transmission area.

(9) Weapons system operation and deployment (FM 3-04.140, applicable weapon system manuals, and unit SOP). Topics in this subject area are—

- Weapons initialization, arming and safety.
- Operation and function of the installed weapon systems.
- Visual search and target detection.
- Duties of the door gunner(s).
- Fire and employment techniques.
- Weapons employment during night and NVD operations.

(10) NVG operations (FM 3-04.140, FM 3-04.203, TC 3-04.93, TM 11-5855-263-10, NVG TSP, and unit SOP). Topics in this subject area are—

- Nomenclature, characteristics, limitations, and operations.
- Mission planning.
- Effects on distance estimation and depth perception.
- Tactical operations, including lighting.
- Use of internal and external lights.
- Terrain interpretation, map preparation, and navigation.

(11) ME and MP system topics: aircraft systems, avionics, mission equipment description and operation, systems malfunctions analysis and troubleshooting. (The flight manual and maintenance manuals may be accessed at https://upw.jtdi.mil.) Topics in this subject area are (for MEs and MPs only)—

- Local airspace usage.
- MTF weather requirements.
- MTF forms and records.
- Electrical system.
- APU.
- Power plant.
- Power train.
- Flight controls.
- Fuel systems.
- Instrument indications.
- Maintenance operations checks (MOCs)/MTF requirements.
- Communications and navigation equipment.
- Caution and Warning Lights.
- Instrument indications.
- Hydraulic/pneumatic systems.
- Vibrations.
- Auto-pilot checks.

(12) SP, IP, IE, UT, SI and FI evaluator/trainer topics (TC 3-04.11 and IP handbook). Topics in this subject area are—

- Learning process.
- Effective communication.
- Teaching methods.
- Flight instruction techniques.
- Human behavior.
- Teaching process.
- Critique and evaluation.
- Effective questions.

c. **Phase 3–Flight Evaluation.** If this phase is required, the following procedures apply.

(1) Briefing. The evaluator will explain the flight evaluation procedure and brief the examinee in the tasks to be evaluated. When evaluating an evaluator/trainer, the evaluator must advise the examinee that, during role-reversal, they may deliberately perform some tasks outside standards to check the examinee's diagnostic and corrective action skills. The evaluator will conduct or have the examinee conduct a crew briefing IAW Task 1000 and the unit's approved aircrew briefing checklist (CL).

(2) Preventive maintenance daily (PMD), pre-flight inspection, engine-start, run-up procedures, engine ground operations, and before-takeoff checks. The evaluator will evaluate the examinee's use of the flight manual and/or the integrated electronic technical manual. The evaluator will have the examinee identify and discuss the function of at least two aircraft systems.

(3) Flight tasks. At a minimum, the evaluator will evaluate those tasks designated by this ATM, tasks listed on the CTL as mandatory for the designated crew station(s) for the type of evaluation being conducted, and those mission/additional tasks selected by the commander. In addition to the commander's selected tasks, the evaluator may evaluate any task performed during the evaluation as long as the task is listed on the crewmember's CTL. Evaluators/trainers must demonstrate an ability to instruct/evaluate flight tasks.

Note. During instrument evaluation, the aviator's vision will be restricted by wearing a vision-limiting device.

(4) Engine shutdown and after-landing tasks. The evaluator will evaluate the examinee's use of the flight manual.

d. **Phase 4–Debriefing.** During this phase, the evaluator will—

(1) Discuss the examinee's strengths and weaknesses.

(2) Offer recommendations for improvement.

(3) Tell the examinee whether they passed or failed the evaluation and discuss any tasks not performed to standards.

(4) Inform the examinee of any restrictions, limitations, or revocations the evaluator will recommend to the commander following an unsatisfactory evaluation.

(5) Complete the applicable forms and ensure the examinee reviews and initials these forms.

3-5. ADDITIONAL EVALUATIONS.

a. **CBRNE evaluation.** This evaluation is conducted IAW TC 3-04.11.

b. **Gunnery evaluation.** This evaluation is conducted IAW FM 3-04.140, the applicable weapons system manual, and the unit SOP.

c. **No-notice, post-mishap flight evaluations and medical flight evaluations.** These evaluations will be conducted IAW AR 95-1.

This page intentionally left blank.

Chapter 4

Crewmember Tasks

> This chapter implements portions of STANAG 3114 AMD Edition 8.

This chapter describes the tasks essential for maintaining crewmember skills. It defines the task title, number, conditions, and standards by which performance is measured. A description of crew actions, along with training and evaluation requirements, is also provided. It does not contain all maneuvers that can be performed in the aircraft.

4-1. TASK CONTENTS.

a. **Task number.** Each ATM task is identified by a 10-digit systems approach to training number. The first three digits of each task in this ATM are 011 (U.S. Army Aviation School); the second three digits are 217. For convenience, only the last four digits are listed in this training circular. The last four digits are as follows:

- Base tasks are assigned 1000-series numbers.
- Mission tasks are assigned 2000-series numbers.
- Additional tasks are assigned 3000-series numbers.
- Maintenance tasks are assigned 4000-series numbers.

Note. Additional tasks designated by the commander as mission essential are not included in this ATM. The commander will develop conditions, standards, and descriptions for those additional tasks.

b. **Task title.** The task title identifies a clearly defined and measurable activity. Titles may be the same in several ATMs, but tasks may be written differently for the specific airframe.

c. **Conditions.** The conditions specify the common conditions under which the task will be performed. Reference will be made to a particular helicopter within a design series when necessary. All conditions must be met before task iterations can be credited.

(1) Common conditions are—

(a) In a mission aircraft with mission equipment and crew, items required by AR 95-1, AR 95-2, FARs, DA Pam 738-751, DOD FLIPs, the flight manual, the commander's ATP, local and unit SOPs.

(b) Under VMC.

(c) Day, night, and NVD employment.

(d) In any terrain or climate.

(e) CBRNE (including mission-oriented protective posture [MOPP]-4) equipment employment.

(f) Electromagnetic environmental effects (E^3).

(2) Common training/evaluation conditions are—

(a) When a SP, IE, IP, UT, or ME is required for training of the task, that individual will be at one set of the flight controls while the training is being conducted. References to the IP in the task conditions include the SP. References to FI in the task conditions include the SI. Evaluators/trainers who are evaluating/training NCMs (except those performing FE duties) must be at a station without access to the flight controls except when evaluating crew coordination.

(b) The following tasks require a SP, IE, or IP for training/evaluation in the aircraft with access to the flight controls. If the IE is not an IP or SP, the IE may only perform the engine failure emergency procedure and Task 1182, and must be trained and evaluated by an SP or IP on the following tasks:

- Task 1070, Respond to Emergencies.
- Task 1074, Respond to Engine Failure at Cruise Flight.
- Task 1075, Perform Single Engine Landing.
- Task 1182, Perform Unusual Attitude Recovery.

(c) Unless otherwise specified in the conditions, all in-flight training/evaluations will be conducted under VMC. IMC denotes flight solely by reference to flight instruments while wearing a vision-limiting device.

(d) Unless specified in the task considerations, a task may be performed in any mode of flight without modifying the standards or descriptions. When personal equipment (NVG, CBRNE, MOPP-4) or mission equipment (water bucket and rescue hoist) is required for task performance, the availability of the equipment becomes part of the conditions.

(e) The aircrew will not attempt the tasks listed below when performance planning indicates OGE power is not available:

- Task 1170, Perform Instrument Takeoff.
- Task 2026, Perform Terrain Flight.
- Task 2036, Perform Terrain Flight Deceleration.
- Task 2048, Perform External (Sling) Load Operations.
- Task 2125, Perform Pinnacle/Ridgeline Operations.
- Task 2127, Perform Combat Maneuvering Flight.
- Any task requiring hovering flight in OGE conditions.

(f) The following emergency procedures cannot be performed in the aircraft except in an actual emergency:

- Touchdown autorotation.
- Single-engine takeoff from the ground.
- Actual engine stoppage in flight or while taxiing.
- Both engine condition levers are out of the detent position while taxiing/flying.
- Jettison of external (sling) load.
- Dual generator failure.
- Two rectifier failures.
- Auto-pilot-OFF external (sling) load hook-up.
- Auto-pilot-OFF combat maneuvering flight.

d. **Standards.** The standards describe the minimum degree of proficiency or standard of performance to which the task must be accomplished. The terms "without error," "properly," and "correctly" apply to all standards. Standards are based on ideal conditions, and many standards are common to several tasks. Individual trainer, instructor, or evaluator pilot techniques are not standards nor used as grading elements. Alternate or additional standards will be listed in individual tasks. Standards unique to the training environment for conditions are established in the training considerations section of each task. Unless otherwise specified in the individual task, the following common standards apply:

(1) All tasks.

(a) Do not exceed aircraft limitations (FE will monitor engine, transmission, blade pitch angle, engine pressure ratio [EPR], N_R, and engine speed (N_G).

(b) Perform crew coordination actions IAW chapter 6 of this ATM.

(2) Takeoff. When taking off from unimproved surfaces, the cabin CE will call the aircraft altitude from the ground to 10 feet in 1-foot increments, and the P/FE will call the aircraft altitude above highest obstacle (AHO) at 25, 50, 75, and 100 feet.

(3) Hover.

(a) Maintain heading, ±10 degrees.

(b) Maintain altitude, ±3 feet.

    (c)   Do not allow drift to exceed 5 feet.

    (d)   Maintain a constant rate of movement appropriate for existing conditions.

    (e)   Maintain ground track with minimum drift.

    (f)   NCM(s) will announce all drift/altitude changes.

(4) In-flight.

    (a)   Maintain heading, ±10 degrees.

    (b)   Maintain altitude, ±100 feet.

    (c)   Maintain airspeed, ±10 knots indicated airspeed (KIAS).

    (d)   Maintain rate of climb or descent, ±200 feet per minute (FPM).

    (e)   Maintain the aircraft in trim.

(5) Approach.

    (a)   When approaching unimproved surfaces, the P/FE will call the aircraft altitude AHO at 100, 75, 50, 25, and 10 feet.

    (b)   When landing on unimproved surfaces, the cabin NCM will call the aircraft altitude from 10 feet to the ground in 1-foot increments.

(6) All tasks with the APU/engines operating (RCMs and NCMs).

    (a)   Maintain airspace surveillance (Task 1026).

    (b)   Apply appropriate environmental considerations.

    (c)   Perform crew coordination actions IAW chapter 6 of this ATM.

e.   Description. The description explains the preferred method for accomplishing the task to meet the standards. This manual cannot address all situations and alternate procedures may be required. Other techniques may be used, as long as the task is accomplished safely and the standards are met. The description applies in all modes of flight during day, night, IMC, NVG, or CBRNE operations. When specific crew actions are required, the task will be broken down into crew actions and procedures as follows:

(1) Crew actions. These define the portions of a task preformed by each crewmember to ensure safe, efficient, and effective task execution. The designations "pilot on the controls (P*)", and "pilot not on the controls (P)" do not refer to PC duties. When required, PC responsibilities are specified. For all tasks, the following responsibilities apply:

    (a)   All crewmembers perform aircrew coordination actions: announce malfunctions or emergency conditions and monitor engines/systems operations, and avionics (navigation/communication), as necessary. During VMC, crewmembers will focus attention primarily outside the aircraft, maintain airspace surveillance, and clear the aircraft. All crewmembers will provide timely warning of traffic and obstacles by announcing the type of hazard, direction, distance and altitude (relative to the aircraft). Crewmembers also announce when attention is focused inside the aircraft (except for momentary scans) and when attention is focused outside the aircraft.

    (b)   PC. The PC is responsible for conducting the mission and operating, securing, and servicing the aircraft they command. The PC ensures a crew briefing is accomplished and the mission is performed IAW the mission briefing, air traffic control (ATC) Federal Aviation administration (FAA) instructions (FAA Order 7110.65R), regulations, and SOP requirements.

    (c)   The PI/FE/CE is responsible for completing tasks as assigned by the PC.

    (d)   P*. The P* is responsible for aircraft control, obstacle avoidance, and proper execution of emergency procedures. The P* will announce any deviation and the reason, from instructions issued. The P* will announce changes in altitude, attitude, airspeed, or direction.

    (e)   P. The P is responsible for navigation, in-flight computations, emergency procedures, and assisting the P* as requested. When duties permit, the P assists the P* with obstacle clearance.

    (f)   CE. The CE is responsible for maintaining airspace surveillance, traffic and obstacle avoidance, safety/security of passengers and equipment, and properly executing emergency procedures. The CE provides assistance to the P* and P as required. He or she is also responsible for the maintenance of their assigned aircraft.

(g)   FE. The FE is responsible for maintaining airspace surveillance, traffic and obstacle avoidance, safety/security of passengers and equipment, and properly executing emergency procedures. They are also responsible for pre-flight, prestart, and engine run-up through engine shutdown checks. They provide assistance to the P* and P as required. They are also responsible for the maintenance of their assigned aircraft.

(2) Procedures. This section explains the portions of a task accomplished by an individual or crew.

f.   **Considerations**. This section defines consideration for task accomplishment under various flight modes (for example, night, NVG, environmental conditions, snow/sand/dust, and mountain/pinnacle/ridgeline operations). Crewmembers must consider additional aspects of a task when performing it in different environmental conditions. The inclusion of environmental considerations in a task does not relieve the commander of the requirement for developing an environmental training program IAW TC 3-04.11. Specific requirements for different aircraft or mission equipment may also be addressed as a consideration. Training considerations establish specific actions and standards used in the training environment.

(1) Night and NVG. Wires and other hazards are much more difficult to detect and must be accurately marked and plotted on maps. Use proper scanning techniques to detect traffic and obstacles and avoid spatial disorientation. The P/FE should make all internal checks (such as computations and frequency changes). Visual barriers (so difficult to view that a determination cannot be made, whether or not they contain barriers or obstacles) will be treated as physical obstacles. Altitude and ground speed are difficult to detect; therefore, artificial illumination may be necessary. Determine the need for artificial lighting before descending below barriers. Adjust search/landing light for best illumination angle without causing excessive reflection into the cockpit. Entering IMC with artificial illumination may induce spatial disorientation. Cockpit controls will be more difficult to locate and identify; take special precautions to identify and confirm the correct switches and levers.

(2) Night unaided. Use of white light or weapons flash will impair night vision. The P* should not directly view white lights, weapons flash, or impact. Allow time for adapting to dark or, if necessary, adjust altitude and airspeed until adapted. Exercise added caution if performing flight tasks before reaching full dark adaptation. Dimly visible objects may be more easily detected using peripheral vision and may tend to disappear when viewed directly. Use off-center viewing techniques to locate and orient on objects.

(3) NVG. Use of NVG degrades distance estimation and depth perception. Aircraft in-flight may appear closer than they actually are due to the amplification of external lights and the lack of background objects to assist in distance estimation and depth perception. If possible, confirm the distance unaided. Weapons flash may temporarily impair or shut down NVG.

g.   **Training and evaluation requirements.** Training and evaluation requirements define whether the task will be trained or evaluated in the aircraft, FS, or academic environment. Listing aircraft/FS under the evaluation requirements does not preclude the evaluator from evaluating elements of the task academically to determine depth of understanding or planning processes. Some task procedures allow multiple ways to achieve the standards.

h.   **References.** References are sources of information relating to a particular task. Certain references apply to many tasks. In addition to the references listed with each task, the following common references apply as indicated.

(1) OEM and vendor manuals that are specific to the operation and maintenance of the Mi-17.

(a)   Mi-17-1V Helicopter Flight Manual.

(b)   Mi-17-1V Helicopter Maintenance Manual–General with Change 1.

(c)   Mi-17-1V Helicopter Maintenance Manual–Airframe.

(d)   Mi-17-1V Helicopter Maintenance Manual–Power Plant.

(e)   Mi-17-1V Helicopter Maintenance Manual–Helicopter Systems.

(f)   Mi-17-1V Helicopter Maintenance Manual–Helicopter Equipment.

(g)   Mi-17-1V Helicopter Maintenance Manual–Pyrotechnic Devices and Aerial Delivery Equipment.

(h)   Mi-17-1V Fire Extinguishers Maintenance Schedule.

(i)   Mi-17-1V Helicopter Maintenance Schedule–Airframe, Helicopter Systems, Power Plant.

    (j)   Mi-17-1V Helicopter Maintenance Schedule–Avionics.

    (k)   Mi-17-1V Helicopter Maintenance Schedule–Helicopter Equipment.

    (l)   Mi-17-1V OEM/Vendor Maintenance Manuals.

    (m) Mi-17-1V Helicopter Standard Specification.

(2) All flight tasks (tasks with APU/engines operating).

    (a)   AR 95-1.

    (b)   FM 1-230.

    (c)   FM 3-04.203.

    (d)   TC 3-04.93.

    (e)   DOD FLIPs.

    (f)   FARs/host country regulations.

    (g)   Unit/local SOPs.

    (h)   Aircraft logbook.

(3) All instrument tasks.

    (a)   AR 95-1.

    (b)   FM 3-04.240.

    (c)   FAA-H-8083-15.

    (d)   DOD FLIPs.

    (e)   Aeronautical Information Manual (AIM).

(4) All tasks with environmental considerations are addressed in FM 3-04.203.

(5) All tasks used in a tactical situation.

    (a)   FM 3-04.111.

    (b)   FM 3-04.113.

    (c)   FM 3-04.140.

    (d)   FM 3-04.203.

4-2. **TASKS.**

a. **Standards versus descriptions.** The standards describe the minimum degree of proficiency/standard of performance to which the task must be accomplished. Attention to the use of "will," "should," "shall," "must." "may," or "can" throughout the text of a task standard is crucial. The description explains one or more recommended techniques for accomplishing the task to meet the standards.

b. **Critical task list.** The following numbered tasks are Mi-17 crewmember critical tasks.

TASK 1000

Participate in a Crew Mission Briefing

CONDITIONS: Prior to flight in a Mi-17 helicopter or a Mi-17 FS and given DA Form 5484 (Mission Schedule/Brief) information and an unit-approved crew briefing CL.

STANDARDS: Appropriate common standards and the following additions/modifications:

1. The PC will acknowledge an understanding of DA Form 5484 and will actively participate the in crew mission brief.

2. The PC will conduct or supervise an aircrew mission briefing (table 4-1) or a more detailed unit-approved crew briefing CL.

3. Crewmembers will verbally acknowledge a complete understanding of the aircrew mission briefing.

DESCRIPTION:

1. Crew actions.

   a. A designated briefing officer will evaluate and brief essential areas of the mission to the PC IAW AR 95-1. The PC will acknowledge a complete understanding of the mission brief and initial DA Form 5484.

   b. The PC has overall responsibility for the crew mission briefing. The PC may direct other crewmembers to perform all or part of the brief.

   c. Crewmembers will direct their attention to the crewmember conducting the briefing. They will address any questions to the briefer and acknowledge understanding of the assigned actions, duties and responsibilities. Lessons learned from previous debriefings should be addressed during the crew briefing, as applicable. If two or more NCMs will perform flight duties, the NCM in charge will be designated on the DA Form 5484 and will brief other cabin NCMs accordingly on their individual responsibilities.

   Note. An inherent element of the crew mission briefing is establishing the time and location for the crew-level after action review (AAR). (Refer to Task 1262.)

2. Procedures.

   a. Brief the mission using a unit-approved aircrew mission briefing CL (table 4-1).

Table 4-1. Sample aircrew briefing checklist

1. Crew introduction/qualifications.
2. Required items: publications, identification tags, ALSE, personnel and mission equipment.
3. Analysis of aircraft.
4. Mission overview, flight route, time line, and notices to airme (NOTAMs).
5. Weather (departure, en route, destination, and void time).
6. Formation/multi-aircraft operations.
7. Tactical considerations; rules of engagement (ROE); weapon engagement rules; weapon status; identification, friend or foe; combat search a d rescue (CSAR) terms: and evasion plan.
8. External (sling) load operations.
9. Airspace surveillance procedures/visual sector duties (Task 1026).
   a. Logbook and pre-flight deficin ies.
   b. Performance planning. Re-computation of performance planning card (PPC).
   c. Mission deviation required based on ircraft analysis.
10. Crew actions, duties, and e pon ibilties.
    a. Transfer of flight controls and two-challenge rule.
    b. Emergency actions.
    (1) Actions to be performed by the P*, P, FE, and NCM.
    (2) Emergency equipment/first aid kits/survival kits.
    (3) Emergency procedures and rendezvous points.

Table 4-1. Sample aircrew briefing checklist (cont.)

(4) IIMC, NVG failure.

(5) Mission consideration. Threat situation, emergency squawk/communication, zeroize equipment, disable aircraft, collect/destroy classified materials, weapons security.

11. General crew duties.

a. P*.

(1) Fly the aircraft–primary focus outside when VMC, inside when IMC.

(2) Avoid traffic and obstacles.

(3) Crosscheck systems and instruments.

(4) Monitor/transmit on radios as directed by the PC.

b. P.

(1) Assist in traffic and obstacle avoidance.

(2) Tune radios and set transponder.

(3) Navigate.

(4) Copy clearances, automatic terminal information services, and other information.

(5) Crosscheck systems and instruments.

(6) Monitor/transmit on radios as directed by the PC.

(7) Read and complete CL items as required.

(8) Announce when focused inside.

c. FE.

(1) Monitor systems and instruments.

(2) Read and complete CL items as required.

(3) Announce when focused outside.

(4) Perform other duties as directed by PC.

d. CE and other assigned crewmembers.

(1) Complete passenger briefing.

(2) Secure passengers and cargo.

(3) Assist in traffic and obstacle avoidance.

(4) Perform other duties as assigned by the PC.

12. Crew-level AAR–time and location.

13. Crewmembers' questions, comments, and acknowledgment of mission briefing.

Note. A safety harness will be worn and secured when performing crew duties. A seat belt will be worn at all times when seated unless it interferes with crew duties.

b.   Table 4-2, page 4-8, provides a suggested format for the minimum mandatory nonrated crewmember briefing CL. Crewmembers will identify mission and flight requirements demanding effective communication and proper sequencing and timing of actions.

Table 4-2. Sample nonrated crewmember briefing checklist

Nonrated crewmember briefing checklist
1. Mission overview.
2. Aircraft run-up responsibilities.
a. Left NCM responsibilities.
b. Right NCM responsibilities.
3. Required items, mission equipment, and personnel
4. Crew actions, duties, and responsibilities.
a. Sectors of responsibility–assist in traffic and obstacle avoidance.
b. Emergency actions.
(1) Mission considerations.
(2) Emergency action with external load.
(3) Egress and rendezvous points.
(4) Actions performed by the FE and CE.
c. Perform other duties assigned by the PC.
e. Refueling operations.
5. Tactical flight.
a. Terrain flight duties.
b. Landing area reconnaissance.
c. Slope.
d. External load procedures.
6. Shut down procedures.
7. Post flight procedures.
8. NCM questions, comments, and acknowledgement of NCM mission briefing.

TRAINING AND EVALUATION REQUIREMENTS:

1. Training will be conducted academically.
2. Evaluation will be conducted academically.

REFERENCES: Appropriate common references, FM 3-04.300, and DA Form 5484.

TASK 1002

Conduct Passenger Briefing

CONDITIONS: Before flight in a Mi-17 helicopter with current mission information.

STANDARDS: Appropriate common standards and the following additions/modifications:

1. The PC will conduct or direct another crewmember to conduct the passenger briefing according to the operator's manual/CL and the unit SOP.

2. Briefer will ensure passengers verbally acknowledge a complete understanding of the passenger briefing.

DESCRIPTION:

1. Crew actions.

   a. The PC has overall responsibility for the passenger briefing, but may direct other crewmembers to perform all or part of it.

   b. Briefer commands attention of passengers. Briefer will require each passenger to verbally confirm understanding of the briefing.

2. Procedures. Examples of passenger briefing items are as follows:

   a. Proper direction to approach and depart the aircraft.

   b. Location of emergency entrances, exits, and equipment.

   c. Use of seat belts.

   d. Location of survival equipment.

   e. Security of equipment.

   f. Actions in the event of an emergency.

TRAINING AND EVALUATION REQUIREMENTS:

1. Training may be conducted in the aircraft or academically.

2. Evaluation may be conducted in the aircraft or academically.

REFERENCES: Appropriate common references and the unit SOP.

TASK 1004

Plan a Visual Flight Rules Flight

CONDITIONS: Prior to VFR flight in a Mi-17 helicopter or a Mi-17 FS, given access to weather information (such as NOTAMs; flight planning aids; necessary charts, forms, and publications; and weight and balance information).

STANDARDS: Appropriate common standards and the following additions/modifications:
1   Determine if the aircrew and aircraft are capable of completing the assigned mission.
2.   Determine if the flight can be performed under VFR.
3.   Determine the correct departure, en route, and destination procedures.
4   Select route(s) and altitudes that avoid hazardous weather conditions. **Do not** exceed aircraft or equipment limitations; if appropriate, select altitudes conforming to VFR cruising altitudes.
5.   Determine distance ±1 nautical mile (NM), ground speed ±5 knots, and estimated time en route (ETE) ±2 minutes for each leg of the flight. Compute magnetic headings ±5 degrees.
6.   Determine the fuel required for the mission, ±100 liters.
7.   Verify the aircraft will remain within weight (WT) and center of gravity (CG) limitations for the duration of the flight IAW the flight manual.
8.   Verify aircraft performance data and ensure power is available to complete the mission IAW the flight manual.
9.   Complete the flight plan.
10. Perform mission risk assessment IAW unit SOP.

DESCRIPTION:
1.   Crew actions.
   a.   The PC may direct the other RCM/FE to complete some elements of the VFR flight planning.
   b.   The other RCM/FE will complete the assigned elements and report the results to the PC.
   c.   The PC will ensure all crewmembers are current and qualified and the aircraft is properly equipped to accomplish the assigned mission.
2.   Procedures.
   a.   Using U.S. military, FAA, or host-country weather facilities, obtain information about the weather. After ensuring the flight can be completed under VFR IAW AR 95-1, check NOTAMs, chart update manuals, and other appropriate sources for any restrictions applying to the flight. Obtain navigational charts covering the entire flight area and allow for any required changes in routing due to weather/terrain.
   b.   Select the course(s) and altitude(s) that will best facilitate mission accomplishment.
   c.   Determine the magnetic heading, ground speed, and ETE for each leg, to include the alternate airfield if required. Compute total distance and flight time, and calculate the required fuel using a CPU-26A/P computer/Weems plotter (or equivalent) or the appropriate Mi-17 charts. Determine if the duplicate weight and balance forms in the aircraft logbook apply to the mission IAW AR 95-1. Verify that aircraft WT and CG will remain within allowable limits for the entire flight. Complete the flight plan and file it with the appropriate agency.

NIGHT OR NIGHT VISION GOGGLE CONSIDERATIONS: More detailed planning is necessary at night due to visibility restrictions. Checkpoints used during the day may not be suitable for night or NVG use.

TRAINING AND EVALUATION REQUIREMENTS:
1.   Training will be conducted academically.
2.   Evaluation will be conducted academically.

REFERENCES: Appropriate common references and aircraft logbook.

TASK 1006

Plan an Instrument Flight Rules Flight

CONDITIONS: Prior to IFR flight in a Mi-17 helicopter or a Mi-17 FS, given access to weather information (such as NOTAMs; flight planning aids; necessary charts, forms, and publications; and weight and balance information).

STANDARDS: Appropriate common standards and the following additions/modifications:

1. Determine if the aircrew and aircraft are capable of completing the assigned mission.

2. Determine if the flight can be performed under IFR IAW AR 95-1, applicable FARs/host nation regulations, local regulations, and SOPs.

3. Determine the proper departure, en route, and destination procedures.

4. Select route(s) and altitude(s) that avoid hazardous weather conditions, best ensure mission completion without exceeding aircraft or equipment limitations, and conform to IFR cruising altitudes. If off-airway, determine the course(s) ±5 degrees and determine the off-airway altitude without error.

5. Select an approach compatible with the weather, approach facilities, and aircraft equipment. Determine if an alternate airfield is required IAW AR 95-1, applicable FARs/host-nation regulations, local regulations, and SOPs.

6. Determine distance ±1 NM, true airspeed ±5 knots, ground speed ±5 knots, and ETE ±2 minutes for each leg of the flight.

7. Determine the fuel required for the mission, ±100 liters.

8. Verify aircraft will remain within WT and CG limitations for the duration of the flight IAW the flight manual.

9. Verify aircraft performance data and ensure power is available to complete the mission IAW the flight manual.

10. Complete and file the flight plan.

11. Perform mission risk assessment IAW unit SOP.

DESCRIPTION:

1. Crew actions.
   a. The PC may direct the other RCM/FE to complete some elements of the IFR flight planning.
   b. The other RCM/FE will complete the assigned elements and report the results to the PC.
   c. The PC will ensure all crewmembers are current and qualified, and the aircraft is properly equipped to accomplish the assigned mission.

2. Procedures.
   a. Obtain weather information using U.S. military, FAA, or host-country weather facilities.
   b. Compare destination forecast and approach minimums, and determine if an alternate airfield is required.
   c. Ensure the flight can be completed IAW AR 95-1.
   d. Check NOTAMs and other appropriate sources for any restrictions applying to the flight.
   e. Obtain navigation charts covering the entire flight area and allowing for routing or destination changes due to weather conditions.
   f. Select the route(s), course(s), and altitudes best facilitating mission accomplishment.
   g. When possible, select preferred routing.
   h. Determine the magnetic heading, ground speed, and ETE for each leg, to include flight to the alternate airfield if required.
   i. Compute the total distance and flight time, and calculate the required fuel using a CPU-26A/P and a computer/Weems plotter (or equivalent) or mission planning system.
   j. Determine if the weight and balance forms in the aircraft logbook apply to the mission IAW AR 95-1.
   k. Verify aircraft WT and CG will remain within allowable limits for the entire flight.
   l. Complete the flight plan and file with the appropriate agency.

TRAINING AND EVALUATION REQUIREMENTS:

1. Training will be conducted academically.
2. Evaluation will be conducted academically.

REFERENCES: Appropriate common references.

TASK 1010

Prepare a Performance Planning Card

CONDITIONS: Given the aircraft takeoff gross weight (GWT), environmental conditions at departure, cruise, and arrival, the flight manual, and a blank DA Form 5701-17 (Mi-17 Performance Planning Card) (figure 4-1, page 4-15).

STANDARDS: Appropriate common standards and compute performance planning data using the flight manual and any approved supplemental data to the flight manual.

DESCRIPTION:

1. Crew duties. The pilot in command (PC) will compute or direct other RCM or the FE to compute the aircraft performance data required to complete the mission. The PC will verify the accuracy of the computations, and ensure that aircraft performance meets mission requirements and aircraft limitations will not be exceeded. If the premission planning figures exceed these values, the mission profile will be reconfigured. This does not preclude flight within additional time-limited operations as limited by the flight manual for events such as unforecasted environmental conditions or unplanned mission requirements.

2. Procedures.

 a. Determine and have available aircraft performance data required to complete the mission. DA Form 5701-17 may be used to aid in organizing performance planning data required for the mission. This form will be used for readiness level (RL) progression training, APART evaluations, and during other training and evaluations when required.

 b. Arrival data is not required to be completed when computing the PPC if environmental data at the destination or intermediate stops has not significantly changed.

 c. When significant changes in the mission's environmental conditions occur, re-compute all affected values. The crew will perform additional hover power checks and re-compute all PPC values any time the environmental conditions change significantly. A significant change is defined as ±1,000 feet pressure altitude (PA), and/or an increase of 10 degrees Celsius. Additionally, an increase of 500 pounds GWT from the departure data will constitute a significant change.

Note. Use mission forecast conditions to obtain the most accurate performance data.

Note. When computing IGE/OGE hover GWTs, adjust the weight by using the graphs in section 1 of the flight manual.

Note. Flight manual sections 1, 3, 4, and 6 contain examples for using the performance data charts. When an example is cited in this description, refer to the appropriate example in these sections.

Note. If any computed value exceeds operating limitations, enter not applicable. Additionally, leave value blank when it does not apply.

Figure 4-1. Sample DA Form 5701-17

3. **DA Form 5701**-17. (All chart/table references in Items 1 to 16 are found in the Kazan Flight Manual.)

   a. **Departure Data.**

      Item 1–PRESSURE ALT. Record the PA forecast for the time of departure.

      Item 2–FAT. Record the free air temperature (FAT) forecast for the time of departure.

      Item 3–WIND. Record the current wind direction and speed for the time of departure.

      Item 4–WEIGHTS.

(a) OPERATING: record the operating weight of the aircraft.

(b) FUEL: record the takeoff fuel weight. If the internal fuel tank is installed add the weight of fuel to the aircraft total weight.

(c) PAX: record the maximum anticipated weight of the number of passengers during the mission profile.

(d) LOAD: record the maximum anticipated weight of the load during the mission profile.

(e) TAKEOFF: record the takeoff gross weight.

Item 5–START.

(a) APU AIR LINE PRESSURE. Using the APU air bleed line pressure chart, record the air pressure for engine start using current PA and FAT temperature. Enter the chart at the bottom using current PA. Move up vertically to the current temperature. Continue left to read air pressure.

(b) MAX power turbine inlet temperature (PTIT) for Start. Using the Gas Generator Idle Speed and Maximum PTIT chart, enter the chart at the bottom using the current temperature, move up vertically until intersecting the PTIT limit line. Move left to record the maximum PTIT for engine start.

(c) ENGINE IDLE (gas producer [speed] [N_1]). Using the same chart, record the minimum and maximum engine idle speed. Enter the chart at the bottom using the current temperature. Move up vertically until intersecting the minimum idle speed line. Move left and record the PTIT. Continue left and record the minimum idle speed. Re-enter the chart at the bottom using current temperature and move up vertically until intersecting the maximum idle speed line. Move left and record the PTIT. Continue left and record the maximum idle speed.

(d) PTIT–Partial acceleration check. Using the Maximum PTIT Versus Ambient Temperature chart for partial acceleration test, enter the chart at the bottom using current temperature. Move up vertically until intersecting the diagonal line. Move left to record the maximum PTIT temperature for the test.

Item 6–HOVER.

Note. The takeoff weight shall be reduced by 200 kilograms (440 pounds) with the DPD ejector (dust protectors) on.

Note. The takeoff weight shall be reduced by 800 kilograms (1764 pounds) with engine and rotor anti-icing on.

Note. Takeoff gross weight can be increased or decreased by using the wind speed and direction chart. Enter the chart on the left side at "0." Move right to the current wind speed. Move either up or down to intercept the headwind, left or right crosswind, or tailwind. Then move left to determine the increase or decrease of takeoff gross weight. At no time will the maximum takeoff gross weight of 13,000 kilograms (28,660 pounds) be exceeded.

(a) IGE. Using the Maximum Takeoff Weight at Vertical Takeoff and Landing IGE chart, enter on the left side at the current temperature. Move right until intersecting the correct altitude line. Move down to read takeoff weight.

(b) OGE. Using the Maximum Takeoff Weight at Vertical Takeoff and Landing Out of Ground Effect chart, enter on the left side at the current temperature. Move right until intersecting the correct altitude line. Move down to read takeoff weight.

b. **Cruise Data.** Enter the forecasted PA, FAT temperature, and Wind for your anticipated en route altitude and flight route. Use the chart for computing Cruise II, Cruise I, Max Continuous, Take-Off Power, and Contingency Power.

Item 7–CRUISE II. Provides minimum fuel consumption per hour (maximum endurance). Use the gas generator speed versus engine inlet air temperature chart. Enter the chart at the bottom using the appropriate temperature. Move vertically until intercepting the minimum Cruise II line. Move left to record the minimum N_G (N_1) . Using the same temperature line again move up until reaching the

Cruise II maximum line. Move left to record the maximum N_G (N_1). Main rotor revolutions per minute (RPM) must be 95, ±2 percent. Oil pressure should be 3 to 4 kilograms fluid per centimeter squared. Maximum PTIT should not exceed 870 degrees Celsius. Time of continuous operation is not limited.

Item 8–CRUISE I. Provides minimum fuel consumption per kilometer (maximum range). Use the same chart as Cruise II. Use the same procedure as Cruise II. Main rotor RPM must be 95, ±2 percent. Oil pressure should be 3 to 4 kilograms fluid per centimeter squared. Maximum PTIT should not exceed 910 degrees Celsius. Time of continuous operation is not limited.

Item 9–MAX CONTINUOUS (Normal power maximum speed). Corresponds to 80 percent of maximum rating power. This power rating can be used during takeoff, climb, and horizontal flight is necessary. Use the same procedure as in Cruise II and I. Main rotor RPM must be 95, ±2 percent. Oil pressure should be 3 to 4 kilograms fluid per centimeter squared. Maximum PTIT should not exceed 955 degrees Celsius. Time of continuous operation is limited to 60 minutes.

Item 10–TAKE-OFF POWER. The rating can be used in case of an engine failure in flight, during engine run up on the ground, and after replacement of the EEC. Use the same procedure as above in Item 9. Main rotor RPM must be 93, ±1 percent. Oil pressure should be 3 to 4 kilograms fluid per centimeter squared. Maximum PTIT should not exceed 990 degrees Celsius. N_G (N_1) at 101.05 percent. Time of continuous operation should not exceed 6 minutes dual engine and 30 minutes single engine.

Item 11–CONTINGENCY POWER. This rating is a single engine operating condition only to be used during actual emergency conditions. Use the same chart and procedures as outlined above. Main rotor RPM must be 93, ±1 percent. Oil pressure should be 3 to 4 kilograms fluid per centimeter squared. Maximum PTIT should not exceed 990 degree Celsius. N_G (N_1) at 98 to 101.15 percent. Time of continuous operation should not exceed 2.5 minutes.

Item 12–AIRSPEED. Use the airspeed limitations table (Kazan Flight Manual, paragraph 2.4) to determine climb, horizontal flight, power-on, and autorotational glide within the speed range indicated on this table.

Item 13/14–OEI (SINGLE Engine). Using the Maximum Helicopter Weight in Horizontal Flight with One Engine Inoperative chart, determine the safe aircraft weight and altitude should one engine fail in flight. Enter the chart at the bottom using the appropriate temperature. Move up vertically until intersecting the correct altitude for flight. Move left to record the maximum aircraft weight during single engine operation.

c. **Arrival Data.** Enter the forecasted PA, FAT Temperature, and Wind for your anticipated landing area. This data is not required if the environmental data at destination or intermediate stops has not significantly changed (±1,000 feet PA, and/or ±10 degrees Celsius) or had an increase of 500 pounds GWT from the departure data.

Item 15–LDG GWT. Record the estimated landing GWT.

Item 16–HOVER.

Note. The landing weight shall be reduced by 200 kilograms (440 pounds) with the DPD ejector (dust protectors) on.

Note. The landing weight shall be reduced by 800 kilograms (1764 pounds) with engine and rotor anti-icing on.

Note. The landing gross weight can be increased or decreased by using the wind speed and direction chart. Enter the chart on the left side at "0." Move right to the current wind speed. Move either up or down to intercept the headwind, left or right crosswind, or tailwind. Move left to determine the increase or decrease of landing gross weight.

(a) IGE. Using the Maximum Takeoff Weight at Vertical Takeoff and Landing IGE chart, enter on the left side at the current temperature. Move right until intersecting the correct altitude line. Move down to read landing weight.

(b) OGE. Using the Maximum Takeoff Weight at Vertical Takeoff and Landing Out of Ground Effect chart, enter on the left side at the current temperature. Move right until intersecting the correct altitude line. Move down to read landing weight.

TRAINING AND EVALUATION REQUIREMENTS:

1. Training will be conducted academically.
2. Evaluation will be conducted academically, in a Mi-17 FS, or in the aircraft.

REFERENCES: Appropriate common references.

TASK 1012

Verify Aircraft Weight and Balance

CONDITIONS: Given crew WTs, aircraft configuration, mission cargo, passenger data, the flight manual, and a completed Department of Defense (DD) Form 365-4 (Weight and Balance Clearance Form F-Tactical/Transport).

STANDARDS: Appropriate common standards and the following additions/modifications:

1. Verify CG and GWT remain within aircraft limits for the duration of the flight IAW the flight manual.

2. Identify all mission or flight limitations imposed by WT or CG.

3. Ensure DD Form 365-4 has been completed within the preceding 90 days.

DESCRIPTION:

1. Crew actions.
   a. The PC will brief crewmembers on any limitations.
   b. Crewmembers will continually monitor aircraft loading during the mission (such as fuel transfers, external [sling] loads, and internal loads) to ensure CG remains within limits.

2. Procedures.
   a. Using the completed DD Form 365-4, verify aircraft GWT and CG will remain within the allowable limits for the entire flight. Note all GWT and/or loading task/maneuver restrictions/aircraft limitations. If there is no completed DD Form 365-4 that meets the mission requirements, the PC will ensure adjustments are made to the existing DD Form 365-4 (to meet the criteria outlined in AR 95-1) and the aircraft is capable of completing the assigned mission.
   b. Verify the aircraft CG in relation to CG limits at predetermined times during the flight when an aircraft's configuration requires special attention (for example, when it is a critical requirement to keep a certain amount of fuel in a particular tank). Conduct CG checks for fuel transfer, external (sling) loads, and cargo loading operations.

TRAINING AND EVALUATION REQUIREMENTS:

1. Training will be conducted academically.

2. Evaluation will be conducted academically.

REFERENCES: Appropriate common references, TM 55-1500-342-23, and DD Form 365-4.

TASK 1013

Operate Mission Planning System

CONDITIONS: In a Mi-17 helicopter or a Mi-17 FS and given an approved computer with mission planning software, a mission briefing, signal operating instructions (SOI) information, weather information, navigational maps, DOD FLIP, intelligence data, and other materials as required.

STANDARDS: Appropriate common standards and the following additions/modifications:

1. Configure and operate the approved mission planning software.
2. Evaluate and enter all pertinent weather data, as appropriate.
3. Select and enter appropriate primary and alternate routes, if required.
4. Select and enter appropriate tactical/terrain flight mission planning control features.
5. Select and enter appropriate communication data.
6. Load mission data to data transfer device, if applicable.
7. Print out time distance heading (TDH) cards, waypoint lists, crew cards, communication cards, and knee-cards as required.

DESCRIPTION:

1. Crew actions. The PC will assign tasks. The crew receives the mission briefing. Any crewmember may enter data into the approved mission planning software and brief the crew on the mission.
2. Procedures. Plan the flight IAW Tasks 1004 or 1006 as applicable, using all appropriate data.

TRAINING AND EVALUATION REQUIREMENTS:

1. Training will be conducted academically.
2. Evaluation will be conducted academically.

REFERENCES: Appropriate common references.

TASK 1014

Operate Aviation Life Support Equipment

CONDITIONS: Given the appropriate ALSE for the mission.

STANDARDS: Appropriate common standards and the following additions/modifications:

1. Inspect/perform operational checks on ALSE.
2. Use personal and mission ALSE IAW the flight manual/instructions for each piece of equipment.
3. Brief/assist passengers in the use of ALSE.

DESCRIPTION:

1. Crew actions. The PC will verify that all required ALSE equipment is onboard the aircraft and meets all serviceability criteria before takeoff.

2. Procedures. Based on mission requirements, obtain the required ALSE. Inspect equipment for serviceability and perform required operational checks. The NCM will secure the required ALSE in the aircraft IAW AR 95-1, the flight manual, and appropriate SOP. The NCM will brief passengers on ALSE use.

TRAINING AND EVALUATION REQUIREMENTS:

1. Training will be conducted in the aircraft or academically.
2. Evaluation will be conducted in the aircraft or academically.

REFERENCES: Appropriate common references, TM 1-1680-377-13&P, , TM 55-1680-317-23&P, and TM 55-1680-351-10.

TASK 1016

Perform Internal Load Operations

CONDITIONS: In a Mi-17 helicopter loaded with passengers/cargo or academically.

STANDARDS: Appropriate common standards and the following additions/modifications:

1. RCM.

    a. Perform or ensure a thorough passenger briefing has been conducted and a passenger manifest is on file, if applicable IAW AR 95-1. Conduct a passenger briefing (Task 1002) IAW the flight manual/CL and the unit SOP.

    b. Verify the aircraft will remain within GWT and CG limitations.

    c. Ensure passengers and cargo are properly restrained.

    d. Ensure floor-loading limits are not exceeded.

    e. Ensure cargo meets restraint criteria.

2. NCM.

    a. Perform a thorough passenger briefing and ensure a passenger manifest is on file, if applicable.

    b. Conduct the briefing IAW the flight manual/CL and unit SOP.

    c. Verify that the aircraft will remain within GWT and CG limitations.

    d. Load the aircraft IAW the load plan, if applicable.

    e. Ensure floor-loading limits are not exceeded.

    f. Secure passengers and cargo.

    g. Ensure cargo meets restraint criteria.

DESCRIPTION:

1. Crew actions.

    a. The PC (with NCM assistance) will formulate a load plan. The PC will ensure a DD Form 365-4 is verified (if required) and the aircraft will be within GWT and CG limits. The PC will ensure the crew loads the cargo, uses proper tie-down procedures, and any passengers receive a briefing. The PC will determine whether the aircraft is capable of completing the assigned mission and ensure aircraft limitations will not be exceeded.

    b. The P* will perform a hover power check before takeoff and ensure the maximum allowable GWT of aircraft is not exceeded.

    c. The NCM will ensure passengers are seated and wearing seat belts before takeoff IAW AR 95-1. The NCM will monitor passengers/cargo during the flight for security.

2. Procedures.

    a. Load cargo IAW the cargo plan or DD Form 365-4. Properly secure and restrain all cargo IAW criteria in the appropriate manuals.

    b. Brief passengers for the flight and seat them IAW the load plan or DD Form 365-4. Conduct the passenger briefing IAW the flight manual/CL or unit SOP and information about the mission. Ensure passengers understand each element of the briefing.

Note. If the aircraft is not shut down for loading, a passenger briefing may be impractical. Passengers may be pre-briefed or passenger briefing cards used IAW the appropriate local directives or unit SOP.

Note. If the cargo/rescue winch is used, the NCM must ensure it is correctly operated IAW the flight manual.

Note. Hazardous cargo will be handled, loaded, and transported IAW AR 95-27 and the flight manual/CL.

TRAINING AND EVALUATION REQUIRMENTS:

1. Training will be conducted in the aircraft or academically.
2. Evaluation will be conducted in the aircraft or academically.

REFERENCES: Appropriate common references, AR 95-27, DD Form 365-4, FM 55-450-2, and TM 55-1500-342-23.

TASK 1019

Perform Preventative Maintenance Daily Check

CONDITIONS: Given a Mi-17 helicopter and the PMD CL.

STANDARDS: Appropriate common standards and the following additions/modifications:
1. Correctly check all items IAW PMD CL.
2. Enter necessary information on the appropriate aircraft documentation.

DESCRIPTION:
1. Using PMD CL, conduct a PMD inspection. When conducting the inspection with another NCM, both NCMs will use the appropriate reference.
2. Obtain a fuel sample from each fuel tank and determine if the sample contains any water or foreign matter. Correctly enter required information in the aircraft logbook.

NIGHT OR NIGHT VISION GOGGLE CONSIDERATIONS: If time permits, accomplish the maintenance inspection during daylight hours. During the hours of darkness, use a flashlight with an unfiltered lens to supplement available lighting. Hydraulic leaks, oil leaks, and other defects are difficult to see using a flashlight with a colored lens.

TRAINING AND EVALUATION REQUIREMENTS:
1. Training will be conducted in the aircraft.
2. Evaluation will be conducted in the aircraft.

REFERENCES: Appropriate common references, aircraft logbook, DA Pam 738-751, appropriate aircraft documentation, FM 3-04.203, TC 3-04.7, and PMD CL.

TASK 1020

Prepare Aircraft for Mission

CONDITIONS: In a Mi-17 helicopter and given a warning order or mission briefing and available mission/required equipment.

STANDARDS: Appropriate common standards and the following additions/modifications:
1. Install, secure, inspect, and inventory all mission equipment.
2. Prepare the aircraft for the assigned mission.

DESCRIPTION:
1. Crew actions. The PC will determine the equipment required for the mission. The PC will verify aircraft is prepared for the mission. The NCMs will ensure required mission equipment is installed, secured, inventoried, and operational before flight.

2. Procedures. After receiving a mission briefing, determine the required mission equipment. Ensure equipment is installed, secured, inventoried, and operational before flight. If an airworthiness release (AWR) is required for mission equipment, ensure a current AWR is in the aircraft logbook and all inspections and checks have been completed IAW the AWR. Check the equipment requiring aircraft power for operation IAW procedures in the flight manual/CL or mission equipment manuals.

TRAINING AND EVALUATION REQUIREMENTS:
1. Training will be conducted in the aircraft.
2. Evaluation will be conducted in the aircraft.

REFERENCES: Appropriate common references, aircraft logbook, and the appropriate AWR, if required.

TASK 1022

Perform Pre-Flight Inspection

CONDITIONS: Given a Mi-17 helicopter and the flight manual/CL.

STANDARDS: Appropriate common standards and the following additions/modifications:
1. RCM.
    a.   Perform the pre-flight inspection IAW the flight manual/CL.
    b.   Enter required information on DA Form 2408-12 (Army Aviator's Flight Record), DA Form 2408-13 (Aircraft Status Information Record), DA Form 2408-13-1 (Aircraft Inspection and Maintenance Record) (IAW DA Pam 738-751), and aircraft documentation.
2. NCM. Complete before pre-flight and pre-flight duties IAW the flight manual/CL and unit SOP, for the designated duty position.

DESCRIPTION:
1. Crew actions.
    a.   The PC will ensure a pre-flight inspection is conducted using the flight manual/CL. The PC may direct other crewmembers to complete elements of the pre-flight inspection as applicable, and will verify all checks have been completed IAW the flight manual/CL. The PC will expediently report aircraft discrepancies that may affect the mission. The PC will ensure the necessary information is entered on the appropriate forms (IAW DA Pam 738-751) and aircraft documentation as required.
    b.   The crewmembers will complete the assigned elements and report to the PC.
2. Procedures.
    a.   The NCM will ensure the aircraft is prepared for pre-flight. The NCM will ensure the aircraft is properly serviced, special equipment is installed, entries in the aircraft logbook are current and correct, and covers and tie-downs are removed. The PC will verify that all pre-flight checks have been completed and ensure that the crewmembers enter the required information on the appropriate forms. The NCM will secure all pre-loaded cargo.
    b.   The NCM will inform the PC of aircraft status to include any special mission equipment installed and all known deficiencies.
    c.   The RCM will use the flight manual to verify each pre-flight check. The NCM will accompany the RCM during the pre-flight inspection, time permitting, and answer each question concerning aircraft components or systems based on data in the aircraft. The NCM will request maintenance assistance as determined necessary by the PC.
    d.   If circumstances permit, accomplish pre-flight inspection during daylight hours.
    e.   The NCM will secure cowlings and equipment following completion of the pre-flight inspection.

NIGHT OR NIGHT VISION GOGGLE CONSIDERATIONS: If performing the pre-flight inspection during the hours of darkness, a flashlight (with an unfiltered lens) should be used to supplement available lighting. Hydraulic leaks, oil leaks, and other defects are difficult to see using a flashlight with a colored lens. Ensure internal and external lighting is operational. FM 3-04.203 contains details regarding nighttime pre-flight inspection.

TRAINING AND EVALUATION REQUIREMENTS:
1. Training will be conducted at the aircraft.
2. Evaluation will be conducted at the aircraft.

REFERENCES: Appropriate common references, FM 3-04.203, DA Pam 738-751, and appropriate aircraft documentation as required.

TASK 1024

Perform Before-Starting-Engine Through Before-Leaving Helicopter Checks

Note. The NCM's visor will be down, unless using NVG, any time a crewmember is outside the aircraft or inside the aircraft with the engines operating, the cabin door is open, and the clamshell doors or the right side escape panel are removed.

CONDITIONS: In a Mi-17 helicopter or a Mi-17 FS and given the flight manual/CL.

STANDARDS: Appropriate common standards and the following additions/modifications:

1. Perform procedures and checks IAW the flight manual/CL.

2. Comply with call and response terminology IAW chapter 6 and the unit SOP.

3. Enter required information on DA Form 2408-12, DA Form 2408-13, DA Form 2408-13-1, and aircraft documentation.

4. Properly secure the aircraft after the last flight of the day IAW the flight manual.

DESCRIPTION:

1. Crew actions.
   a. Each crewmember will complete the required checks pertaining to their assigned crew duties IAW the flight manual/CL. Crewmembers will coordinate with each other before entering data into aircraft systems.
   b. The PC/PI will read the CL and announce APU and engine starts.
   c. All crewmembers will clear the area around the aircraft before APU start and each engine start.
   d. NCMs will perform duties as required by their duty position and those directed by the PC, IAW the unit SOP, while maintaining situational awareness (SA).
   e. The PC will ensure the required information is entered on DA Form 2408-12, DA Form 2408-13, DA Form 2408-13-1 (IAW DA Pam 738-751), and aircraft documentation (IAW approved publications).
   f. Secure the aircraft after completing the flight IAW the flight manual and unit SOP.

2. Procedures.
   a. Perform the before-starting-engine through before-leaving-helicopter checks IAW the flight manual/CL. The call and response method should be used, as appropriate.
   b. The crewmember reading the CL will read the complete CL item.
   c. The crewmember performing the check will answer with the appropriate response. For example, for the call "Anti-collision/position lights–As required," the response might be "Anti-collision lights, both, night; position lights, steady, bright." "As required" is not an appropriate response. Responses that do not clearly communicate action of information should not be used. For example, when responding to the call, "Systems–Check" replying with "Check" does not clearly indicate the systems are within the normal operating range. A response of "All in the normal operating range" communicates more accurate information.
   d. After the flight, enter all information required on the appropriate DA forms or aircraft documentation.
   e. During APU start, an NCM will be outside the aircraft to ensure the area is clear and perform fireguard duties.
   f. During engine start, an NCM will assume a position so the pilots can see them. The NCM must be standing inside the rotor disk to ensure that the aircraft is clear and ready for the engine start.
   g. The NCM/PIs will complete the post flight IAW the flight manual/CL.
   h. On completion of required maintenance and inspection, the NCM will verify the aircraft is properly moored and protective covers and security devices are properly installed IAW PM-NSRWA manuals and the applicable flight manual.
   i. Perform additional security duties as directed by the PC.

NIGHT OR NIGHT VISION GOGGLE CONSIDERATIONS: Prior to starting the engines, ensure internal and external lights are operational and set. Internal lighting levels must be high enough to easily see the instruments and start the engines without exceeding operating limitations.

SNOW/SAND/DUST CONSIDERATIONS: Ensure all rotating components and inlets/exhausts are clear of ice/snow before starting APU/engines.

TRAINING AND EVALUATION REQUIREMENTS:
1. Training may be conducted in the aircraft or a Mi-17 FS.
2. Evaluation will be conducted in the aircraft or a Mi-17 FS.

REFERENCES: Appropriate common references, DA Pam 738-751, and appropriate aircraft documentation as required.

TASK 1026

Maintain Airspace Surveillance

> **WARNING**
>
> During flight and taxi, the NCMs must be secured to a tie-down fitting in the cabin area. NCMs <u>will not</u> secure their restraining harness to the clamshell door.

CONDITIONS: In a Mi-17 helicopter or a Mi-17 FS in VMC.

STANDARDS: Appropriate common standards and the following additions/modifications:

1. Brief airspace surveillance procedures prior to flight; this will include scan sectors.

2. Announce any unplanned drift or altitude changes; clear the aircraft and immediately inform other crewmembers of all air traffic or obstacles posing a threat to the aircraft.

3. Announce when attention is focused inside the aircraft.

4. Maintain airspace surveillance in assigned scan sectors.

5. When landing, the crew will confirm the suitability of the area and that the aircraft is clear of obstacles.

DESCRIPTION:

1. Crew actions.

    a. The PC will brief airspace surveillance procedures prior to the flight; this briefing will include areas of responsibility and scan sectors.

    b. The P* will announce the intent to perform a specific maneuver and remain focused outside the aircraft. The P* is responsible for clearing the aircraft and obstacle avoidance.

    c. The P and NCM, as duties permit, will assist in clearing the aircraft and provide adequate warning of obstacles, unusual drift, or altitude changes. The P and NCM will announce when their attention is focused inside the aircraft, and again, when attention is reestablished outside.

    d. When landing, the crew will confirm the suitability of the area and that the aircraft is clear of barriers/obstacles. The NCMs will move about the aircraft as necessary to ensure total coverage.

2. Procedures.

    a. Maintain close surveillance of the surrounding airspace.

    (1) Keep the aircraft clear from other aircraft and obstacles by maintaining visual surveillance (close, mid, and far areas) of the surrounding airspace.

    (2) Inform the crew immediately of air traffic or obstacles posing a threat to the aircraft. Call out the location of traffic or obstacles by the clock, altitude, and distance method. (The 12 o'clock position is at the nose of the aircraft.)

    (3) Give distance in miles or fractions of miles for air traffic and in feet for ground obstacles, as appropriate. When reporting air traffic, specify (if known) the type (fixed-wing or helicopter) and model of the aircraft. The altitude of the air traffic should be reported as the same altitude, higher, or lower than the altitude at which you are flying.

    b. Prior to changing altitude, visually and verbally clear the aircraft for hazards and obstacles, inclusive of what is ahead, above, below, to the left, and to the right of the aircraft.

    c. Prior to performing a descending flight maneuver, it may sometimes be desirable to perform clearing "S" turns to the left or right. The clearing "S" turns will provide the aircrew with a greater visual scan area.

    d. During a hover or hovering flight, inform the P* of any unannounced drift or altitude changes. When landing, the crew will confirm the suitability of the area.

NIGHT OR NIGHT VISION GOGGLE CONSIDERATIONS: The use of proper scanning techniques will assist in detecting traffic and obstacles and avoiding spatial disorientation. Hazards such as wires are difficult to detect.

TRAINING AND EVALUATION REQUIREMENTS:

1. Training may be conducted in the aircraft or a Mi-17 FS.
2. Evaluation will be conducted in the aircraft.

REFERENCES: Appropriate common references.

TASK 1028

Perform Hover/Power Check

CONDITIONS: In a Mi-17 helicopter or a Mi-17 FS, at appropriate hover height, with performance planning information available, and the before-hover check completed.

STANDARDS: Appropriate common standards and the following additions/modifications:

1. Position the aircraft in the vicinity of the takeoff point and the direction of takeoff at the appropriate hover height.

2. Determine if aircraft performance is sufficient to complete the mission by noting the engine N_1, PTITs, engine pressure ratio (EPR) indications, pitch setting, and rotor speed (N_R).

3. Determine if sufficient fuel is available to complete the mission.

4. Determine controllability, CG change, and proper instrument response.

5. Determine if aircraft performance is sufficient to complete the mission.

DESCRIPTION:

1. Crew actions.

   a. The PC will determine whether the aircraft is capable of completing the assigned mission and ensure aircraft limitations are not exceeded.

   b. The P* will announce his intent to bring the aircraft to a stationary hover for the hover power check.

   (1) During the ascent, check for proper CG and control response.

   (2) Remain focused outside the aircraft and announce when the aircraft is stabilized at the desired hover altitude.

   (3) Using a 10-foot main gear height, hover near the takeoff point and in the direction of takeoff when performing the hover power check, unless mission or terrain constraints dictate otherwise.

   c. During the ascent, the P/FE will monitor the instruments and verify the power check.

   (1) The P/FE will ensure aircraft limitations are not exceeded and note the N_1, PTITs, EPR indications, pitch setting, N_R, engine oil temperatures, engine oil pressures, and transmission temperature and pressure. The P/FE should determine if sufficient fuel is available to complete the mission.

   (2) The P/FE will announce results to the P*.

   d. Cabin NCMs will remain focused primarily outside the aircraft to assist in clearing and provide adequate warning of obstacles.

2. Procedures.

   a. While hovering, check for proper control response by applying aft, then forward cyclic followed by left and right cyclic. Attitude indicator will indicate proper attitude changes. Apply left, then right pedal, changing heading 3 to 5 degrees.

   b. Note proper changes to heading indicators, magnetic compass, turn needle, and trim ball is free in race.

   c. Slightly increase and decrease collective and note changes in radar and barometric altimeters and proper response from the vertical speed indicators. While at a stabilized hover, check control positioning to ensure proper CG and control rigging.

   d. The PC will ensure aircraft performance and fuel are sufficient to complete the mission.

NIGHT OR NIGHT VISION GOGGLE CONSIDERATIONS: The crew will use proper scanning techniques to avoid excessive drift when hovering at night or using NVG.

TRAINING AND EVALUATION REQUIREMENTS:

1. Training may be conducted in the aircraft or a Mi-17 FS.
2. Evaluation will be conducted in the aircraft or a Mi-17 FS.

REFERENCES: Appropriate common references.

TASK 1032

Perform Radio Communication Procedures

CONDITIONS: In a Mi-17 helicopter or a Mi-17 FS.

STANDARDS: Appropriate common standards and the following additions/modifications:

1. RCMs.
   a. Program, check, and operate aircraft avionics.
   b. Establish radio contact with the desired unit or ATC facility. When communicating with ATC facilities, use radio communication procedures and phraseology outlined in the AIM, DOD FLIPs, and Department of Transportation (DOT)/FAA Order 7110.65 P/CG.
   c. Operate the intercom system.
   d. Perform or describe two-way radio failure procedures IAW the DOD FLIP or host country regulations.
2. NCMs. The internal communication system is used to communicate with the crew.

DESCRIPTION:

1. Crew actions.
   a. The PC will determine radio frequencies IAW mission requirements during the crew briefing and indicate whether the P* or P will establish and maintain primary communications.
   b. The P* will announce information not monitored by the P.
   c. The P/FE will adjust avionics to required frequencies. The P/FE will copy pertinent information and announce information not monitored by the P*.
   d. During normal operations, the NCM will monitor external communications to prevent interruption when external communications are transmitted or received. (Monitoring external communications may not be desirable during operations requiring extensive internal communication, such as external loads, hoist, or emergencies.)
   e. Two-way radio failure:

      (1) P* or P will announce two-way radio failure to all crewmembers.

      (2) The PC will direct efforts to identify and correct the avionics malfunction.

      (3) The P* will focus outside the aircraft during VMC or inside on the instruments during IMC, as appropriate, but should not participate in troubleshooting the malfunction.

      (4) The P will remain focused primarily inside the aircraft to identify and correct the avionics malfunction. The FE must remain primarily focused outside the aircraft in order to maintain the P scan.

   f. Aircraft intercom failure:

      (1) The PC will direct assistance from the crew to determine the malfunction and corrective action. Alternate actions may include switching to a different internal communication system (ICS) box, changing microphone cords if available, hooking up to a different ICS station, hand and arm signals, or passing notes.

      (2) If the problem cannot be corrected, the PC will determine the course of action, which may vary from landing as soon as practical to landing as soon as possible.

2. Procedures.
   a. Adjust avionics to the required frequencies. Continuously monitor the avionics as directed by the PC.

      (1) When required, establish communications with the desired facility.

      (2) Monitor the frequency before transmitting. Transmit the desired/required information. Use the correct radio callsign when acknowledging each communication.

      (3) When advised to change frequencies, acknowledge instructions. Select the new frequency as soon as possible unless instructed to do so at a specific time, fix, or altitude.

      (4) Use radio communication procedures and phraseology as appropriate for the area of operations. Use standard terms and phraseology for all intercommunications.

  b. Two-way radio failure:

   (1) Attempt to identify and correct the malfunctioning radio and announce the results.

   (2) If two-way radio failure is confirmed, comply with procedures outlined in the flight information handbook or host country regulations.

TRAINING AND EVALUATION REQUIREMENTS:

   1. Training may be conducted in the aircraft or a Mi-17 FS.

   2. Evaluation may be conducted in the aircraft or a Mi-17 FS.

REFERENCES: Appropriate common references, AIMs, DOT/FAA Order 7110.65 P/CG, and appropriate avionics operator's manuals.

TASK 1034

Perform Ground Taxi

CONDITIONS: In a Mi-17 helicopter or a Mi-17 FS on a suitable surface, with the before-taxi/after-landing check completed and the aircraft cleared.

STANDARDS: Appropriate common standards and the following additions/modifications:

1. RCM.
    a. Maintain a constant speed appropriate for conditions and stay within ground control limitations.
    b. Maintain desired ground track.
    c. Maintain flight controls IAW the flight manual.
2. NCM.
    a. Perform applicable checks IAW the flight manual/CL and the unit SOP when read by the P/FE.
    b. Immediately inform the RCMs of any observed discrepancy or malfunction.
    c. Clear the aircraft.
    d. Use hand-and-arm signals, if required, IAW FM 21-60.

DESCRIPTION:

1. The P* will ensure the main-rotor RPM is within limits and the parking brake is released. The P* will announce his or her intent to begin the taxi, state the taxi plan, and clear the aircraft.

2. The P and NCMs will assist the P* in clearing the aircraft.

3. The P* will initiate the taxi by increasing the collective slightly (a 1- to 3-degree pitch) and moving the cyclic slightly forward to start movement. Perform taxi check (brakes, heading, and turn indicators). When the aircraft starts moving, reduce the collective to the minimum required to maintain movement at the desired speed. Control heading with the pedals. Use left and right pedal input to turn the aircraft and a slightly lateral cyclic into turns to maintain a level fuselage attitude (cyclic movement should be minimized to avoid droop-stop pounding). Regulate taxi speed with a combination of cyclic, collective, and necessary brake applications. Soft, rough, or sloping terrain may require the use of more or less power than would normally be required.

> *Note.* The P* may use lateral cyclic inputs to assist with directional control. These inputs are normally required while taxiing in a crosswind. Adhere to crosswind restrictions.

4. When the NCM is required outside the aircraft during taxi, the NCM will be positioned where the P*/P can clearly see all hand-and-arm signals or will remain attached to the aircraft communication system.

> *Note.* Emergency stops may be performed with the wheel brakes or by bringing the aircraft to a hover, depending on ground velocity.

> *Note.* Do not attempt cyclic aerodynamic braking during taxi.

> *Note.* Ground taxi is prohibited with wind speeds in excess of 29 knots (hovering is an alternative).

> *Note.* If helicopter vibration increases (ground resonance) during taxi, immediately reduce collective, center the cyclic, and retard throttle to idle. If vibration persists, perform emergency shutdown (fuel stopcocks closed, fuel pumps off, fuel fire shutoff valves closed).

NIGHT OR NIGHT VISION GOGGLE CONSIDERATIONS: Aircraft taxi at night or with NVG requires a constant scan by every crewmember to ensure obstacle clearance. Use of artificial illumination, such as the white or infrared landing/search lights, taxi light, and/or blade tip lights may be necessary for safe operations as determined by the P*. Position lights; anti-collision light should be set to bright. The P* will utilize a ground guide when taxi is required in a congested area. Taxi speeds may need to be reduced.

SNOW/SAND/DUST CONSIDERATIONS: If ground reference is lost due to blowing snow/sand/dust, lower the collective, neutralize the flight controls, and apply wheel brakes until visual reference is reestablished.

Note. Use caution when taxiing near other maneuvering aircraft because of limited visual references and relative motion illusion.

Note. Due to decreased visual references and possibility of relative motion illusion, limit ground speed to a rate appropriate for conditions.

Note. At night, use of the landing, search, taxi, blade tip, or anti-collision lights may cause spatial disorientation in blowing snow/sand/dust.

TRAINING AND EVALUATION REQUIREMENTS:
1. Training will be conducted at the aircraft or a Mi-17 FS.
2. Evaluation will be conducted at the aircraft or a Mi-17 FS.

REFERENCES: Appropriate common references and FM 21-60.

TASK 1038

Perform Hovering Flight

CONDITIONS: In a Mi-17 helicopter or a Mi-17 FS, with the before-hover check completed and aircraft cleared.

STANDARDS: Appropriate common standards and the following additions/modifications:

1. Perform a smooth, controlled ascent to hover.
2. Perform a smooth, controlled descent with minimal drift at touchdown.
3. Maintain ground track, ±5 feet during hover taxi.
4. Maintain a constant rate of turn, not to exceed 12 degrees per second.

DESCRIPTION:

1. Crew actions.

   a.   The P* will announce his intent to perform a specific hovering flight maneuver and will remain focused primarily outside the aircraft to monitor altitude and avoid obstacles. The P* will ensure and announce the aircraft is cleared prior to turning or repositioning the aircraft. The P* will announce when they terminate the maneuver.

   b.   The P and NCMs will assist in clearing the aircraft in the direction of any movement and provide adequate warning of obstacles, excessive drift, or excessive altitude changes. They will announce when their attention is focused inside the aircraft and again when their attention is reestablished outside.

   c.   The P and NCMs should assist the P* in maintaining the position of the aircraft over the pivot point.

2. Procedures.

   a.   Takeoff to a hover.

   (1)   Position the cyclic, as necessary, and maintain heading with the pedals.

   (2)   Smoothly raise the collective control and adjust the cyclic to ascend vertically to a 10-foot hover altitude, unless mission requirements dictate another altitude.

   (3)   Release the brakes as necessary.

   b.   Hovering flight.

   (1)   Adjust the cyclic to maintain a stationary hover or to hover in the desired direction. Control heading with the pedals and maintain altitude with the collective control.

   (2)   Maintain a constant hover speed appropriate for the conditions. To return to a stationary hover, apply cyclic in the opposite direction while maintaining heading with the pedals and altitude with the collective control.

Note. Air taxi is the preferred method for ground movements on airports, provided ground operations and conditions permit. Unless otherwise requested or instructed, pilots are expected to remain below 100 feet above ground level (AGL). However, if a higher than normal airspeed or altitude is desired, the request should be made prior to lift-off. The pilot is solely responsible for selecting a safe airspeed for the altitude/operation being conducted. Using air taxi enables the pilot to proceed at an optimum airspeed/altitude, minimize downwash effect, conserve fuel, and expedite movement from one point to another.

---

   c.   Hovering turns.

   (1)   Apply pressure to the desired pedal to begin the turn. Use pressure and counter pressure on the pedals to maintain the desired rate of turn. Coordinate cyclic control to maintain position over the pivot point while maintaining altitude with the collective. Hovering turns can be made around any vertical axis (for example, the nose, mast, tail of the aircraft, or a point in front of the aircraft). However, turns other than approximately the center of the aircraft will increase the turn radius proportionately.

(2) Around the nose. With the aircraft stationary, pick a point, slightly forward of the nose. Control the direction and rate of turn with the cyclic and pedals and maintain altitude with the collective control (cross-control of the cyclic and pedals is required to pivot around the nose).

(3) Around the mast. With the aircraft at a stationary hover and the mast over the pivot point, apply pedal in the desired direction of turn. Maintain a stationary position over the pivot point with the cyclic. Control the rate of turn with the pedals and maintain altitude with the collective; this results in the minimum radius turn.

(4) Around the tail. With the aircraft at a stationary hover and the pivot point under the tail, apply cyclic and pedal in the direction of the intended turn. Use cyclic and pedal to control the rate of turn and movement. Maintain hover altitude with the collective.

    d. Landing from a hover.

(1) Lower the collective control to affect a smooth rate of descent until the main gear contacts the ground.

(2) Coordinate collective reduction with the cyclic, as necessary, to maintain pitch attitude and to stop horizontal movement.

(3) Smoothly lower the collective control to allow the nose gear to contact the ground. Continue to lower the collective full down, neutralize the controls, and apply brakes to stop forward movement.

(4) If landing to an improved surface, if clear, allow aircraft to roll forward for proper landing gear positioning.

(5) If sloping conditions are suspected or anticipated, refer to Task 1062.

3. The P and NCMs should assist the P* in maintaining the position of the aircraft over the pivot point.

4. When landing from a hover to an unimproved area, the crew must check for obstacles under the aircraft. Brake may be set prior to touchdown.

> *Note.* Cyclic turns should only be used when necessary.

NIGHT OR NIGHT VISION GOGGLE CONSIDERATIONS:

1. Movement over areas of limited contrast–such as tall grass, water, or desert–tends to cause spatial disorientation. Seek hover areas providing adequate contrast and use proper scanning techniques. If disorientation occurs, apply sufficient power and execute an instrument takeoff (ITO) (Task 1170). If a takeoff is not feasible, try to maneuver the aircraft forward and down to the ground to limit the possibility of touchdown with lateral or aft movement.

2. Night: Unaided hovering flight is difficult due to loss of visual references around the aircraft. If needed, utilize the landing light (white) to help clear the aircraft and to increase awareness of aircraft movement. If hovering is required in a congested area, ground guides will be utilized. Use of the white light will impair night vision for several minutes. Therefore, exercise added caution if resuming flight before reaching full dark adaptation.

NIGHT VISION GOGGLES: Aided hovering flight requires constant scanning and acute awareness of aircraft movement above the ground. While wearing NVG both depth perception and distance estimation capabilities are reduced. This is primarily due to reduced visual acuity, as opposed to day vision, and the lack of peripheral vision cues. Depth perception in a given situation depends on available light, type and quality of the goggles, degree of contrast in the field of view, and the viewer's experience. Increased scanning techniques and a thorough understanding of the monocular cues will help in performing hovering flight. If hovering is required in a congested area, a ground guide will be utilized.

> *Note.* Hovering is permitted in wind speeds (head, tail, or cross) no higher than limits identified in the flight manual.

SNOW/SAND/DUST CONSIDERATIONS:

1. During ascent to a hover, if visual references do not deteriorate to an unacceptable level, continue ascent to desired hover altitude.

2. A 10-foot hover taxi: During takeoff to a hover, simultaneously accelerate the aircraft to a ground speed that keeps the snow/sand/dust cloud just aft of the main rotor mast.

> *Note.* Maintain optimum visibility by observing references close to the aircraft. Exercise caution when operating in close proximity to other aircraft or obstacles.
>
> *Note.* When visual references deteriorate making a 10-foot hover taxi unsafe, determine whether to abort the maneuver, ground taxi, air taxi, or perform an ITO (Task 1170).

3. A 20-to 100-foot air taxi: Use this maneuver when it is necessary to move the aircraft over terrain unsuitable for hover taxi. Initiate air taxi the same as a 10-foot hover, but increase altitude to not more than 100 feet and accelerate to a safe airspeed appropriate for conditions.

> *Note.* Ensure an area is available to safely decelerate and land the aircraft. Under certain conditions, such as adverse winds, it may be necessary to perform a traffic pattern to optimize conditions at the desired termination point.
>
> *Note.* Hovering OGE reduces available ground references and may increase the possibility of spatial disorientation. Be prepared to transition to instruments and execute an ITO (Task 1170) or unusual attitude recovery (Task 1182) if ground reference is lost.
>
> *Note.* At night, use of landing, search, or anti-collision lights may cause spatial disorientation while in blowing snow/sand/dust.

CONFINED AREA CONSIDERATIONS: Select good references to avoid unanticipated drift. All crewmembers must be focused primarily outside for obstacle avoidance.

TRAINING AND EVALUATION REQUIREMENTS:

1. Training will be conducted in the aircraft or a Mi-17 FS.
2. Evaluation will be conducted in the aircraft.

REFERENCES: Appropriate common references.

TASK 1040

Perform Visual Meteorological Conditions Takeoff

CONDITIONS: In a Mi-17 helicopter or a Mi-17 FS, with the hover power and before-takeoff checks completed and the aircraft cleared.

STANDARDS: Appropriate common standards and the following additions/modifications:
1. Maintain takeoff heading, ±10 degrees below 50 feet AGL.
2. Maintain ground track aligned with takeoff direction.
3. Maintain aircraft in trim above 50 feet AGL.
4. Maintain takeoff power until reaching desired climb airspeed, or transition to mission profile.

DESCRIPTION:
1. Crew actions.
   a.   The PC will determine the direction of takeoff by analyzing the tactical situation, the wind, the long axis of the takeoff area, and the lowest obstacles, and will confirm required power is available. The PC will ensure the required fuel for the mission is available or add sufficient fuel, abort, or revise the mission.
   b.   The P* will remain focused primarily outside the aircraft throughout the maneuver to provide obstacle clearance. The P* will announce whether the takeoff is from the ground or from a hover and their intent to abort or alter the takeoff. The P* will select reference points to assist in maintaining the takeoff flight path.
   c.   The P and NCMs will announce when ready for takeoff and will remain focused primarily outside the aircraft, to assist in clearing and to provide adequate warning of obstacles.

   Note. The cabin NCM will verify that passengers, cargo, and mission equipment are properly secured.

---

   d.   The P/FE will monitor power requirements and advise the P* if power limits are being approached. The P/FE will announce when their attention is focused inside the aircraft and again when attention is reestablished outside.
2. Procedures.
   a.   From the ground.

   (1) The P* will announce his or her intent to take off from the ground. The P* will focus his or her attention primarily outside the aircraft but will occasionally cross-check the flight instruments.

   (2) The P and NCMs will announce when ready for takeoff and will remain focused primarily outside the aircraft, to assist in clearing and to provide adequate warning of obstacles.

   (3) The P* will select reference points to maintain ground track. With the cyclic and pedals in the neutral position, the P* will release the brakes and raise the collective control until the aircraft is airborne and accelerating.

   (4) All landing gear should leave the ground at the same time. As the aircraft leaves the ground, the P* will apply forward cyclic as required to smoothly accelerate through effective translational lift (ETL) at an altitude appropriate for the terrain and obstacles.

   (5) The P* will adjust the cyclic as necessary to continue the acceleration (approximately 5 degrees nose down, not to exceed 10 degrees nose down), obtain the desired climb airspeed, and maintain ground track. The P* will position the collective control as necessary to clear obstacles in the flight path and obtain the desired rate of climb. The P* will use the pedals to maintain heading when below 50 feet AGL and in trim when above 50 feet AGL.

   (6) When the P* obtains the desired climb airspeed, he or she will adjust the cyclic as necessary to stop the acceleration. The P* will adjust the collective to continue or to stop the rate of climb.

(7) The P/FE will monitor power requirements and advise the P* if power limits are being approached. The P and NCMs will announce when their attention is focused inside the aircraft and again when attention is reestablished outside.

    b.   From a hover.

(1) The P* will announce his or her intent to take off from a hover and will focus his or her attention primarily outside the aircraft.

(2) The P and NCM will announce when ready for takeoff and will remain focused primarily outside the aircraft to assist in clearing and to provide adequate warning of obstacles.

(3) The P* will select reference points to maintain ground track. The P* will apply forward cyclic to smoothly accelerate the aircraft through ETL while adjusting the collective, as required, to maintain the appropriate hover height. The P* will perform the rest of the maneuver similar to a takeoff from the ground.

(4) The P/FE will monitor power requirements and advise the P* if power limits are being approached. The P and NCMs will announce when their attention is focused inside the aircraft and again when attention is reestablished outside.

Note. Performing this maneuver in certain environments may require hover OGE power. Evaluate each situation for power required versus power available, such as terrain flight takeoff.

Note. The P* must avoid excessive and unnecessary nose-low accelerative attitudes.

Note. The NCMs should remain seated during this maneuver.

NIGHT OR NIGHT VISION GOGGLE CONSIDERATIONS:

1. If sufficient illumination exists to view obstacles, the P* should accomplish the takeoff at night similar to a VMC takeoff during daylight. Visual obstacles, such as shadows, should be treated the same as physical obstacles.

2. If sufficient illumination does not exist, perform an altitude-over-airspeed takeoff by applying takeoff power first, followed by a slow acceleration to ensure obstacle clearance. The P* may perform the takeoff from the ground or a hover. Maintain takeoff power setting until reaching climb airspeed. Adjust the collective as required to establish the desired rate of climb and cyclic to maintain the desired airspeed. The P* and cabin NCMs should maintain orientation outside the aircraft and concentrate on obstacle avoidance.

3. The P/FE should make all internal checks. The P/FE will advise the P* when the altimeter and vertical speed indicator (VSI) show climb has been established (phrased "you have a climb outside and inside [altimeters and VSI]). Reduced visual references during takeoff and throughout ascent at night may make it difficult to maintain the desired ground track. Knowledge of surface wind direction and velocity will assist in maintaining the desired ground track. Use proper scanning techniques to avoid spatial disorientation. Ensure the searchlight or landing light (white light) is in the desired position when performing operations during unaided night flight. Using the white light will impair night vision for several minutes. Exercise added caution if resuming flight before reaching full dark adaptation.

SNOW/SAND/DUST CONSIDERATIONS: Apply collective and cyclic as required to ascend vertically. As the aircraft leaves the ground, maintain heading with the pedals and a level attitude with the cyclic. As the aircraft clears the snow/sand/dust cloud and clears the barriers, accelerate to climb airspeed and trim the aircraft.

Note. In some cases, applying the collective to blow away loose snow/sand/dust from around the aircraft is beneficial before performing this maneuver.

Note. At night, use of the landing, search, or anti-collision lights may cause spatial disorientation while in blowing snow/sand/dust.

> *Note.* Be prepared to transition to instruments and execute an instrument takeoff if ground reference is lost.

CONFINED AREA CONSIDERATIONS: Before departure, confirm the takeoff plan. Perform a hover power check, if required. Reposition the aircraft, if desired, to afford a shallower departure angle and minimize power requirements. During departure, adjust the cyclic and the collective as required to establish a constant departure angle to clear obstacles. All crewmembers must be focused primarily outside for obstacle avoidance.

MOUNTAIN/PINNACLE/RIDGELINE CONSIDERATIONS: Analyze winds, obstacles, and density altitude. Perform a hover power check. Determine the best takeoff direction and path for conditions. After clearing any obstacle(s), accelerate the aircraft to the desired airspeed.

> *Note.* Where drop-offs are located along the takeoff path, the aircraft may be maneuvered down slope to gain airspeed.

MUD/MUSKEG/TUNDRA CONSIDERATIONS: Perform one of the following takeoff techniques:

1. Dry muskeg/tundra areas. A vertical takeoff may be best in drier areas where the aircraft has not sunk into the muskeg/tundra or where obstacles prohibit motion. Smoothly increase the collective until the crew confirms that the wheels/skis are free. Adjust controls as necessary to perform a VMC takeoff.

2. Wet areas. In wet areas where the aircraft is likely to have sunk or is stuck in the mud/muskeg/tundra, the following technique may be best: with the cyclic in the neutral position, smoothly increase the collective. As hover power is approached, place the cyclic slightly forward of the neutral position and slowly move the pedals back and forth. Continue increasing the collective and "swim" the aircraft forward to break the suction of the wheels/skis. When free, adjust the controls as necessary to perform a VMC takeoff.

> *Note.* Before performing operations in a mud/muskeg/tundra environment, it is important to understand dynamic rollover characteristics.

TRAINING AND EVALUATION REQUIREMENTS:

1. Training will be conducted in the aircraft or a Mi-17 FS.
2. Evaluation will be conducted the in aircraft.

REFERENCES: Appropriate common references.

TASK 1044

Navigate by Pilotage and Dead Reckoning

CONDITIONS: In a Mi-17 helicopter or a Mi-17 FS and given a plotter, a computer, the flight log, and appropriate maps.

STANDARDS: Appropriate common standards and the following additions/modifications:

1. Maintain orientation, ±500 meters.
2. Arrive at check points/destination at estimated time of arrival (ETA), ±2 minutes.

DESCRIPTION:

1. Crew actions.

   a. The P* will focus primarily outside the aircraft and respond to navigation instructions or cues given by the P. The P* will acknowledge commands issued by the P for the heading, altitude, and airspeed changes necessary to navigate the desire course. The P* will announce significant surface features to assist in navigation.

   b. The P will direct the P* to change aircraft heading, altitude, and airspeed as appropriate to navigate the desired course. The P will use rally terms, specific headings, relative bearings, or key terrain features to accomplish this task. The P will announce all plotted wires before approaching their location. The P/FE will monitor aircraft instruments, assist in clearing the aircraft, and provide adequate warning to avoid traffic and obstacles. The P/FE will announce when their attention is focused inside the aircraft and again when attention is reestablished outside.

   c. The cabin NCMs will continually watch for traffic and obstacles along the flight path.

2. Procedures.

   a. Both pilotage and dead reckoning will be used to maintain the position of the aircraft along the planned route. Planned heading will be adjusted as necessary to compensate for the effects of the wind.

   b. Perform a ground speed check as soon as possible by computing the actual time required to fly a known distance. Adjust estimated time for subsequent legs of the flight route using the computed ground speed. Compare planned ground speed with computed ground speed and adjust airspeed as required to arrive at each control point at its original ETA.

NIGHT OR NIGHT VISION GOGGLE CONSIDERATIONS: More detailed flight planning is required when the flight is conducted at night. Interior cockpit lighting should be considered when selecting colors for preparing navigation aids such as maps and kneeboard notes. FM 3-04.203 contains details on night navigation and mission planning.

TRAINING AND EVALUATION REQUIRMENTS:

1. Training will be conducted in the aircraft or a Mi-17 FS.
2. Evaluation will be conducted in the aircraft.

REFERENCES: Appropriate common references.

TASK 1046

Perform Electronically Aided Navigation

CONDITIONS: In a Mi-17 helicopter or a Mi-17 FS with an electronically aided navigation system installed and operational.

STANDARDS: Appropriate common standards and the following additions/modifications:

1. Operate the installed electronically aided navigation system IAW the appropriate technical manual or manufacturer's operating manual.

2. Determine the position of the aircraft along the route of flight within ±300 meters.

3. Arrive at check points/destination at ETA, ±1 minute.

DESCRIPTION:

1. Crew actions.

   a. The P* will focus primarily outside the aircraft and respond to navigation instructions or cues given by the P. The P* will acknowledge commands issued by the P for the heading, altitude, and airspeed changes necessary to navigate the desire course. The P* will announce significant surface features to assist in navigation.

   b. The P/FE will be primary operator of the electronically-aided navigation system. The P will direct the P* to change aircraft heading, altitude, and airspeed as appropriate to navigate the desired course. The P will use rally terms, specific headings, relative bearings, or key terrain features to accomplish this task. The P will announce all plotted wires before approaching their location.

   Note. Only the P/FE will perform in-flight time/labor intensive navigation programming duties (for example, building routes).

   c. The P/FE will monitor aircraft instruments, assist in clearing the aircraft, and provide adequate warning to avoid traffic and obstacles.

   d. The P and FE will announce when their attention is diverted from their normal duties and again when attention is reestablished.

2. Procedures. Perform the turn-on, test, and programming procedures IAW the appropriate technical manual. The proper updating and shutdown procedures will be performed IAW the appropriate technical manual or flight manual. The P* will use the heading indicators with the global positioning system (GPS) when flying the selected course.

   Note. Use of any "VFR ONLY" GPS system as an IFR navigational system for IMC operations is not authorized; however, the crew should consider and plan for its use as an emergency backup system.

TRAINING AND EVALUATION REQUIRMENTS:

1. Training will be conducted in the aircraft or a Mi-17 FS.

2. Evaluation will be conducted in the aircraft or a Mi-17 FS.

REFERENCES: Appropriate common references.

TASK 1048

Perform Fuel Management Procedures

WARNING

Failure to monitor fuel operations could result in engine flameout because of fuel starvation.

CONDITIONS: In a Mi-17 helicopter or a Mi-17 FS, with a CPU-26A/P computer or calculator.

STANDARDS: Appropriate common standards plus the following additions/modifications:

1. RCM.
   a. Verify the required amount of fuel is onboard at the time of takeoff.
   b. Initiate an alternate course of action if the actual fuel consumption varies from the planned value and the flight cannot be completed without the planned use of the required reserve.

2. FE.
   a. Initiate an in-flight fuel consumption check within 10 minutes of leveling off or within 10 minutes of entering into the mission profile.
   b. Within 15 minutes after taking the initial readings, compute the fuel consumption rate ±50 liters per hour and complete the fuel consumption check.
   c. Monitor the remaining fuel quantity and the continuing rate of consumption.

DESCRIPTION:

1. Crew actions.
   a. The FE will record the initial fuel figures, fuel flow computation, burnout, and reserve times. The FE will announce when initiating the fuel check and when completing the fuel check. The FE also will announce the results of the fuel check.

   Note. The FE will ensure that the fuel quantity selector switch is returned to the "service" position any time after the switch is moved from the "service" position.

   b. The PC will acknowledge the results of the fuel check.
   c. The P will confirm the results of the fuel check

2. Procedures.
   a. When performing the before-takeoff check, determine the total fuel onboard, and compare it with the fuel required for the mission. If the fuel onboard is inadequate, add sufficient fuel or abort/revise the mission.
   b. Initial airborne fuel reading. Within 10 minutes after leveling off or within 10 minutes of entering into the mission profile, record the total fuel quantity and the time of reading. Complete the fuel consumption check 15 minutes after taking the initial airborne fuel reading. Determine whether the remaining fuel is sufficient to complete the flight without the planned use of the required reserve.

   Note. Crews should verify ability to transfer fuel from internal to external tanks before using external tank fuel quantities in fuel reserve/burnout computations.

   Note. Do not perform fuel consumption checks while transferring fuel from internal fuel tanks to external fuel tanks.

   c. Fuel quantity and consumption. Periodically monitor the fuel quantity and consumption rate. If the fuel quantity or flow indicates a deviation from computed values, repeat the fuel consumption check to determine if the amount of fuel is adequate to complete the flight. Periodically check individual fuel tank indicators to determine the system is operating properly.

d.   Auxiliary fuel management. Follow procedures outlined in the flight manual when using the external extended range fuel system. Refer to the flight manual when using nonstandard auxiliary fuel systems.

TRAINING AND EVALUATION REQUIREMENTS:

1.   Training may be conducted in the aircraft or a Mi-17 FS.
2.   Evaluation may be conducted in the aircraft or a Mi-17 FS.

REFERENCES: Appropriate common references.

TASK 1052

Perform Visual Meteorological Conditions Flight Maneuvers

CONDITIONS: In a Mi-17 helicopter or a Mi-17 FS, with the aircraft cleared and given VMC.

STANDARDS: Appropriate common standards and the following additions/modifications:

1. Turns.
    a. Clear the aircraft.
    b. Maintain aircraft in trim.
    c. Maintain selected airspeed, ±10 knots.
    d. Maintain altitude, ±100 feet.
    e. Maintain selected bank angle, ±10 degrees.
    f. Roll out on desired heading, ±10 degrees.

2. Climbs and descents.
    a. Clear the aircraft.
    b. Maintain aircraft in trim.
    c. Maintain selected airspeed, ±10 knots.
    d. Maintain rate of climb or descent, ±200 FPM.
    e. Maintain desired heading, ±10 degrees.

3. Straight and level flight.
    a. Maintain aircraft in trim.
    b. Maintain selected airspeed, ±10 knots.
    c. Maintain altitude, ±100 feet.
    d. Maintain desired heading, ±10 degrees.

4. Traffic pattern flight. Enter, operate in, and depart from a traffic pattern.

DESCRIPTION:

1. Crew actions.
    a. The P* will remain focused primarily outside the aircraft. The P* will announce and clear each turn, climb, and descent.
    b. The P and NCMs will assist in clearing the aircraft and will provide adequate warning of traffic and obstacles. They will announce when their attention is focused inside the aircraft and again when attention is reestablished outside.

2. Procedures. Adjust cyclic as required to maintain the desired airspeed, course, ground track, or heading as appropriate. Adjust collective as required to maintain the desired climb/descent rate or altitude and maintain aircraft in trim with the pedals. Perform traffic pattern operations IAW ATC directives, local SOP, and FM 3-04.203.
    a. VMC climb. The P* will raise the collective lever to initiate climb. The P* will adjust the pedals to maintain aircraft in trim. The P* will lower the collective lever to stop climb at desired altitude.
    b. VMC climbing turns. The P* will raise the collective lever to initiate climb. The P* will adjust the pedals to maintain aircraft in trim and apply cyclic in the desired direction of turn. The P* will adjust the cyclic as required to stop turn on heading and lower the collective lever to stop climb at desired altitude.
    c. VMC straight-and-level flight. The P* will adjust the collective lever to maintain altitude. The will adjust the pedals to maintain aircraft in trim. The P* will maintain airspeed and heading.
    d. VMC level turns. The P* will apply cyclic in the desired direction of turn. The P* will adjust the collective lever to maintain altitude and adjust the pedals to maintain aircraft in trim. The P* will apply cyclic opposite the direction of turn to stop the turn on the desired heading.
    e. VMC descents. The P* will lower the collective lever to initiate the descent. He or she will adjust the pedals to maintain aircraft in trim. The P* will raise the collective lever to stop rate of descent at the desired altitude.
    f. VMC descending turns. The P* will lower the collective lever to initiate descent. The P* will adjust the pedals to maintain aircraft in trim and apply cyclic in the desired direction of turn. The P* will adjust

cyclic as required to stop turn at the desired heading. The P* will raise the collective lever to stop the descent at desired altitude.

g. Traffic pattern flight.

(1) The P* will maneuver the aircraft into position to enter the downwind leg midfield at a 45-degree angle (or IAW local procedures), at traffic pattern altitude, and at the desired airspeed. (A straight-in or base-leg entry may be used if approved by ATC.) On downwind, the P will complete the before-landing check.

(a) Before turning base, the P* will lower the collective lever and adjust airspeed as required and initiate a descent. If performing a straight-in or a base-leg entry, the P* will reduce airspeed at a point to facilitate the approach. The P* will turn base and final leg, as appropriate, to maintain the desired ground track.

(b) The P* will perform the desired approach. The P* will announce each turn in the pattern and the type of approach planned. The P and NCMs will assist in clearing the aircraft throughout each turn in the traffic pattern.

(2) For a closed traffic pattern after takeoff, the P* will climb straight ahead at climb airspeed to the appropriate altitude, turn to crosswind, and continue the climb. The P* will initiate the turn to downwind as required to maintain the desired ground track. The P* will adjust the collective lever and cyclic as required to maintain traffic pattern altitude and airspeed.

h. Before-landing check.

(1) The P/FE will perform the before-landing check before turning base.

(2) The P/FE will call out the before-landing check and announce when it is completed.

NIGHT OR NIGHT VISION GOGGLE CONSIDERATIONS:

1. The P* will focus primarily outside the aircraft and should concentrate on obstacle avoidance and aircraft control. The P/FE will make all internal cockpit checks.

2. For NVG training in a traffic pattern, the recommended maximum airspeed is 100 KIAS, and the maximum bank angle is 30 degrees.

TRAINING CONSIDERATIONS: For traffic pattern training, the—

1. Recommended airspeed is 80 KIAS for crosswind and base legs.

2. Rate of climb/descent on crosswind and base legs is 500 FPM.

3. Downwind leg is 100 KIAS.

TRAINING AND EVALUATION REQUIREMENTS:

1. Training may be conducted in the aircraft or a Mi-17 FS.

2. Evaluation will be conducted in the aircraft.

REFERENCES: Appropriate common references.

TASK 1054

Select Landing Zone/Pickup Zone/Holding Area

WARNING

Not all hazards will be shown on a map. When using a map reconnaissance to determine suitability, the added risk of unknown hazards must be addressed during the mission risk assessment process.

CONDITIONS: In a Mi-17 helicopter or a Mi-17 FS and given a map or photo data.

STANDARDS: Appropriate common standards plus the following additions/modifications:

1. Perform map, photo, or visual reconnaissance.

2. Determine the landing zone (LZ) is suitable for operations and provide accurate and detailed information to the supported unit (if applicable).

3. Confirm suitability on initial approach.

DESCRIPTION:

1. Crew actions. The crew will confirm the location of plotted hazards and call out the location of un-plotted hazards.

    a. The PC will confirm suitability of the area for the planned mission.

    b. The P* will remain focused primarily outside the aircraft throughout the maneuver for aircraft control and obstacle avoidance. The P* will announce his or her intent to deviate from the maneuver.

    c. The P and NCMs will assist in LZ reconnaissance and clearing the aircraft. They will provide adequate warning of obstacles and will acknowledge the P*'s intent to deviate from the maneuver.

2. Procedures. Gather map or photo data on potential LZ or conduct an in-flight suitability check if map or photo data is unreliable. Determine suitability by evaluating size, long axis, barriers, surface conditions, tactical situation, and effects of the wind. Select a flight path, altitude, and airspeed that afford the best observation of the landing area, as required. Determine an approach, desired touchdown point, and departure path. The tactical, technical, and meteorological elements must be considered in determining suitability.

Note. If wind conditions will be a factor, a wind evaluation should be performed. Techniques for evaluating wind conditions are found in FM 3-04.203.

Note. Depending on the mission, an in-flight suitability check may not be feasible. Suitability may be determined by a map reconnaissance. Make a final determination of suitability upon arrival at the LZ/pickup zone (PZ).

---

    a. Tactical.

        (1) Mission. Determine if the mission can be done from the selected LZ. Consider flight time, fuel, number of sorties, and access routes.

        (2) Location. To reduce troop fatigue, consider distance of PZ or LZ from supported unit or objective. Also consider the supported unit's mission, equipment, and method of travel to/from PZ/LZ/holding area.

        (3) Security. Consider size and proximity of threat elements versus availability of security forces. The supported unit normally provides security. Consider cover and concealment, key terrain, avenues of approach and departure. The area should be large enough to provide dispersion.

b. Technical.

(1) Number and type of aircraft. Determine if the size of the LZ can support all the aircraft at once or if they must rotate into the LZ for in-flight linkup.

(2) Landing formation. Plan landing formation for shape and size of the LZ.

(3) External (sling) loads. For missions requiring sling loads at or near maximum GWT of the helicopter, select larger LZs where barriers have minimum vertical development.

(4) Surface conditions. Consider slopes, blowing sand, snow, or dust. Be aware that vegetation may conceal surface hazards (for example, large rocks, ruts, or stumps). Areas selected should also be free of sources of rotor wash signature.

(5) Obstacles. Hazards within the LZ that cannot be eliminated must be plotted. Plan approach and departure routes over lowest obstacles.

c. Meteorological.

(1) Ceiling and visibility. Ceiling and visibility are critical when operating near threat elements. IIMC recovery can expose the aircraft and crew to radar-guided and heat-seeking weapons, with few options for detection and avoidance. If one aircrew of a multi-aircraft operation must respond to IIMC, the element of surprise will be lost, the assets onboard will not be available for the mission, and the entire mission may be at risk.

(2) Winds. Determine approach and departure paths.

(3) Pressure altitude. High PA may limit loads and, therefore, require more sorties.

Note. Avoid planning approach or departure routes into a rising or setting sun or moon.

NIGHT OR NIGHT VISION GOGGLE CONSIDERATIONS:

1. Unimproved and unlit areas are more difficult to evaluate at night because of low contrast. Knowledge of the various methods for determining the height of obstacles is critical to successfully completing this task. Visual obstacles such as shadows should be treated the same as physical obstacles.

2. When performing operations during unaided night flight, ensure the searchlight or landing light (white light) is in the desired position. Using the white light will impair night vision for several minutes; therefore, exercise added caution if resuming flight before reaching full dark adaptation.

CONFINED AREA CONSIDERATIONS: Determine a suitable axis and path for a go-around. For multi-aircraft operations, determine the number of aircraft the area can accommodate safely.

SNOW/SAND/DUST CONSIDERATIONS: Evaluate surface conditions for the likelihood of encountering a whiteout/brownout. Determine a suitable axis and path for a go-around.

MOUNTAIN/PINNACLE/RIDGELINE CONSIDERATIONS: When practical, position the aircraft on the windward side of the area. Evaluate suitability–paying particular attention to PA and winds. Determine a suitable axis and escape route for a go-around. Operations at high altitudes are more likely to expose the crews to visual detection, radar, or heat-seeking weapons.

TRAINING AND EVALUATION REQUIREMENTS:

1. Training may be conducted in the aircraft or a Mi-17 FS.
2. Evaluation will be conducted in the aircraft.

REFERENCES: Appropriate common references.

TASK 1058

Perform Visual Meteorological Conditions Approach

CONDITIONS: In a Mi-17 helicopter or a Mi-17 FS, with the before-landing check completed.

STANDARDS: Appropriate common standards and the following additions/modifications:

1. Verify sufficient power for the approach.

2. Maintain a constant approach angle clear of obstacles to the desired point of termination (hover) or touchdown (surface).

3. Maintain a rate of closure appropriate for the conditions.

4. Maintain ground track alignment with the landing direction, as appropriate.

5. Align aircraft with landing direction below 50 feet AGL or as appropriate for transition from terrain flight.

6. Perform a smooth and controlled termination to a hover or touchdown to the surface.

7. Verify crew, passengers, cargo, and mission equipment are secured.

DESCRIPTION:

1. Crew actions.

    a. The P* will select a suitable landing area (analyze suitability, power available, barriers, winds, approach path, touchdown point, and takeoff directions). The P* will focus his attention primarily outside the aircraft to ensure obstacle clearance throughout the approach and landing. The P* will announce when he or she begins the approach and whether they will terminate the approach to a hover or to the ground. The P* will announce the intended point of landing and any deviation from the approach, to include execution of a go-around if necessary.

    b. The P and NCMs will confirm the suitability of the landing area as requested and will assist the P* in clearing the aircraft and warn of any traffic or obstacles. If a go-around is necessary, the P and NCM will remain focused outside the aircraft for obstacle avoidance. The P* will acknowledge any pertinent observations made during the approach. The P and NCMs will announce when focused inside the aircraft and again when attention is reestablished outside.

2. Procedures. Evaluate winds. Select an approach angle allowing obstacle clearance while descending to the desired point of termination. Once the termination point is sighted and the approach angle is intercepted, adjust collective as necessary to establish and maintain a constant angle. Maintain entry airspeed until the rate of closure appears to be increasing. Above 50 feet AGL, maintain ground track alignment and the aircraft in trim. Below 50 feet AGL, align the aircraft with the landing direction. Progressively decrease the rate of descent and rate of closure until reaching the termination point (hover, touchdown) or until a decision is made to perform a go-around.

    a. To a hover. The approach to a hover may terminate with a full stop over the planned termination point or continue movement to transition to hovering flight. Progressively decrease the rate of descent and rate of closure until an appropriate hover is established over the intended termination point.

    b. To the ground. The decision to terminate to the surface with zero speed or with forward movement will depend on the aircraft's loading/environmental conditions. Touchdown with minimum forward or lateral movement. After ground contact, ensure the aircraft remains stable with all movement stopped. Smoothly lower the collective to full down and neutralize the pedals and cyclic. Apply brakes if required.

    c. Go-around. The P* should perform a go-around if a successful landing is doubtful or if visual reference with the intended termination point is lost. Once climb is established, reassess the situation and develop a new course of action.

    Note. The P* should perform a go-around if a successful landing is doubtful or if visual reference with the intended termination point is lost (Task 1068).

    Note. If wind conditions are perceived to be a factor, a wind evaluation should be performed. Techniques for evaluating wind conditions are found in FM 3-04.203.

Note. Steep approaches can place the aircraft in potential settling with power conditions.

Note. Performing this maneuver in certain environments may require hover OGE power. Evaluate each situation for power required versus power available.

NIGHT OR NIGHT VISION GOGGLE CONSIDERATIONS:

1. Altitude, apparent ground speed, and rate of closure are difficult to estimate at night. The rate of descent during the final 100 feet should be slightly less than during the day to avoid abrupt attitude changes at low altitudes. After establishing the descent during unaided flights, airspeed may be reduced to approximately 50 knots until apparent ground speed and rate of closure appear to be increasing. Progressively decrease the rate of descent and forward speed until termination of maneuver.

2. Surrounding terrain or vegetation may decrease contrast and cause degraded depth perception during the approach. Before descending below obstacles, determine the need for artificial lighting.

3. Use proper scanning techniques to avoid spatial disorientation.

4. When performing operations during unaided night flight, ensure the searchlight or landing light (white light) is in the desired position. Use of the white light will impair night vision for several minutes; therefore, exercise added caution if resuming flight before reaching fully dark adaptation.

SNOW/SAND/DUST CONSIDERATIONS:

1. Termination to a point OGE.
   a. This approach requires OGE power and may be used for most snow landings and some sand/dust landings.
   b. Make the approach to a hover OGE over the intended landing point.
   c. Slowly lower the collective and allow the aircraft to descend. The rate of descent will be determined by the rate in which the snow/sand/dust is blown from the intended landing point.
   d. Remain above the snow/sand/dust cloud until it dissipates and visual references can be seen for touchdown. After ground contact, slowly lower the collective to fully down and neutralize the flight controls.

2. Termination to the surface with forward speed.
   a. This termination may be made to an improved landing surface with minimal ground references.
   b. Once the appropriate approach angle is intercepted, adjust the collective as necessary to establish and maintain the angle.
   c. As apparent rate of closure appears to increase, progressively decrease the rate of descent and rate of closure to arrive at the touchdown area slightly above ETL. At this point, maintain the minimum rate of closure to ensure the snow/sand/dust cloud remains behind the pilot's station.
   d. When the wheels or heels of the skis contact the ground, lower the collective and allow the aircraft to settle. Apply slight aft cyclic at touchdown to prevent burying the wheels or toes of the skis. When the wheels or heels of the skis contact the ground, slowly lower the collective and allow the aircraft to settle. Lower the collective as necessary, neutralize the flight controls, and apply brakes as necessary to stop forward movement.

3. Termination to the surface with no forward speed.
   a. This termination should be made for landing areas where slopes, obstacles, or unfamiliar terrain preclude a landing with forward speed.
   b. It is not recommended when new or powder snow or fine dust is present because white/brown out conditions will occur.
   c. The termination is made directly to a reference point on the ground with no forward speed. After ground contact, smoothly lower the collective to full down position and neutralize the flight controls.

Note. Brakes set or released may be determined by the type of surface, hard or soft, during reconnaissance.

d.   Packed surface area. Thin layer of snow or dust atop hard subsurface with some visible terrain elements, such as rocks. Set the brakes to minimize forward roll after landing.

e.   Soft surface area. Thick layer of snow or dust with no visible subsurface, release the brakes to minimize abrupt stop after landing and unnecessary stress on the aft landing gear.

Note. When landing in deep snow, aircraft wheels/skis may settle at different rates and the aircraft will normally terminate in a tail low attitude. During sand/dust landings, all doors and windows should be closed and vent blowers turned off.

Note. Hovering OGE reduces available ground references and may increase the possibility of spatial disorientation. Be prepared to transition to instruments and execute an instrument takeoff if ground reference is lost.

Note. At night, use of the landing, search, or strobe light may cause spatial disorientation while in blowing snow/sand/dust.

CONFINED AREA CONSIDERATIONS: An approach to the forward one-third of the area will reduce the approach angle and minimize power requirements. Before beginning the approach, the crew will determine and brief an escape route in case a go-around is necessary. During the approach, continue to determine the suitability of the area and the possible need for a go-around. If possible, make the decision to go-around before descending below the barriers or going below ETL. After touching down, check aircraft stability as the collective is lowered.

MOUNTAIN/PINNACLE/RIDGELINE CONSIDERATIONS: Select a shallow to steep approach angle depending on the wind, density altitude, GWT, and obstacles. Before beginning the approach, the crew will determine and brief an escape route in case a go-around is necessary. During the approach, continue to determine the suitability of the intended landing area and the possible need for a go-around. If possible, make the decision to go-around before descending below the barriers or going below ETL. After touchdown, check aircraft suitability as the collective is lowered.

Note. To successfully operate into small areas, it may be necessary to place the nose of the aircraft over the edge of the landing area. This may cause a loss of important visual references when on final approach. Every crewmember will assist in providing information on aircraft position in the landing area.

Note. Motion parallax may make the rate of closure difficult to determine until the aircraft is close to the landing area. Reduce airspeed to slightly above ETL until the rate of closure can be determined.

Note. On approach, avoid descents greater than 300 FPM.

MUD/MUSKEG/TUNDRA CONSIDERATIONS: Select a suitable area and terminate the approach to a 10-foot HOVER over the intended touchdown point. Begin a vertical descent until the aircraft touches down. Verify aircraft stability while lowering the collective. If the area is determined by the P* to be suitable, lower the collective fully down and neutralize the cyclic and pedals.

TRAINING AND EVALUATION REQUIRMENTS:
1.   Training will be conducted in the aircraft or a Mi-17 FS.
2.   Evaluation will be conducted in the aircraft.

REFERENCES: Appropriate common references.

TASK 1062

Perform Slope Operations

CONDITIONS: In a Mi-17 helicopter with aircraft cleared.

STANDARDS: Appropriate common standards and the following additions/modifications:

1. RCM.
   a. Select a suitable landing area (within the allowable slope limits).
   b. Set the parking brakes prior to landing.
   c. Execute a smooth and controlled descent.
   d. Maintain heading, ±5 degrees.
   e. Maintain minimum drift (±1 foot) before touchdown and then no drift allowed after wheel contact.
   f. Execute a smooth, controlled ascent.
   g. Monitor slope angle throughout operation.

2. NCM.
   a. Confirm suitable landing area.
   b. Clear the aircraft throughout the landing and/or sequence.
   c. Monitor slope angle throughout operation.

DESCRIPTION:

1. Crew actions.
   a. The P* will announce his intent to perform a slope operation and will establish the aircraft over the slope. The pilot in the left seat will set the brakes. The P* will remain within slope limitations. The P* will announce his or her intended landing area and any deviation from the intended maneuver. The P* should be aware of the common tendency to become tense and, as a result, to over control the aircraft while performing the slope operation. The P* will note the aircraft attitude at a hover prior to starting descent to land on the slope.

   Note. The Mi-17 wheel brake control lever is installed only on the pilot's cyclic stick (left seat position); therefore, the left seat pilot will set the brakes and the right seat pilot will verify they are set.

   ---

   b. The P and NCMs will confirm the suitability of the intended landing area and provide adequate warning of obstacles, excessive drift, or excessive attitude changes. They will announce when their attention is focused inside the aircraft and again when attention is reestablished outside.
   c. The P/FE will note the aircraft attitude on the attitude indicator and notify the P* prior to exceeding aircraft slope limitations.
   d. The cabin NCMs will assume a position where they can observe the slope operation. The NCM will clear their sector while checking that the rotor blades are clear of obstacles and the ground. The cabin NCM will call out wheel height from 10 feet in 1-foot increments until the landing gear contacts the ground. The NCM will advise the P* when all landing gear are on the ground and the aircraft is stable.

2. Procedures.
   a. Upslope landings.

   (1) With the aircraft heading upslope, the P* will lower the collective until the nose gear contacts the ground, maintain heading with the pedals, and adjust cyclic as necessary to maintain the position of the aircraft. The P* will continue to lower the collective control until the main landing gear contacts the ground.

   (2) When the landing gear are on the ground, the P* will smoothly lower the collective to full down position. The P* will then neutralize the controls while checking the stability of the aircraft.

   (3) The P* will perform takeoff from the upslope in the reverse sequence.

b. Downslope landings.

(1) With the aircraft heading downslope the P* will lower the collective until the main landing gear contacts the ground. The P* will adjust pitch attitude to maintain a stabilized position on the slope by coordinating collective reduction with aft cyclic movement (avoiding droop-stop pounding).

(2) The P* will maintain heading with the pedals and smoothly and continuously lower the collective until the nose gear contacts the ground. If the aircraft slides down the slope, the P* will smoothly and deliberately bring it back to a hover and reposition the aircraft.

(3) When the landing gear are on the ground, the P* will smoothly lower the collective to the full down position. The P* will then neutralize the controls while checking the stability of the aircraft.

(4) The P* will perform takeoff from the downslope in the reverse sequence.

c. Cross-slope landings.

(1) With the aircraft heading cross slope, the P* will lower the collective until the upslope main landing gear contacts the ground. The P* will maintain heading with the cyclic and pedals as required without encountering droop-stop pounding.

(2) The P* will maintain pitch attitude by coordinating collective reduction with aft cyclic movement. This will normally place the downslope main landing gear in contact with the ground. The P* will coordinate the cyclic and pedals as necessary and continue to lower the collective until the nose gear is on the ground.

(3) The P* will smoothly lower the collective to the full down position and neutralize the controls while checking the stability of the aircraft.

(4) The P* will perform the takeoff from the cross slope in the reverse sequence.

d. Takeoff. Before takeoff, announce initiation of an ascent. Smoothly increase the collective and apply the cyclic into the slope to maintain the position of the upslope wheel. Continue to increase the collective to raise the downslope wheel(s), maintain heading with the pedals, and simultaneously adjust the cyclic to attain a hover attitude. As the aircraft leaves the ground, adjust the cyclic to accomplish a vertical ascent to a hover with minimum drift.

Note. Before conducting slope operations, RCMs will understand the characteristics of droop-stop pounding and dynamic rollover.

Note. If at any time successful completion of the landing is in doubt, the P* will abort the maneuver.

Note. Crewmembers must be aware of the helicopter's normal hovering attitude before putting a wheel on the ground.

NIGHT OR NIGHT VISION GOGGLE CONSIDERATIONS: When conducting slope operation, determine the need for artificial illumination before starting the maneuver. Select reference points to determine slope angles. (References probably will be limited and difficult to ascertain.) If at any time, successful completion of the landing is doubtful, abort the maneuver. When performing operations during unaided night flight, ensure the searchlight or landing light (white light) is in the desired position. Using the white light will impair night vision for several minutes; therefore, exercise added caution if resuming flight before reaching fully dark adaptation.

TRAINING AND EVALUATION REQUIREMENTS:

1. Training will be conducted in the aircraft or a Mi-17 FS.
2. Evaluation will be conducted in the aircraft.

REFERENCES: Appropriate common references.

TASK 1064

Perform Roll-On Landing

CONDITIONS: In a Mi-17 helicopter or a Mi-17 FS and given a suitable landing area, with the before-landing check completed.

STANDARDS: Appropriate common standards and the following additions/modifications:
1. Select a suitable landing area.
2. Maintain a constant approach angle clear of obstacles to desire point of touchdown.
3. Maintain a ground track alignment that aligns with the landing direction.
4. Execute a smooth, controlled touchdown at speed appropriate for the conditions, but not exceeding 27 knots ground speed.
5. Touchdown with a maximum of 10 degrees nose high pitch attitude aligned with the landing directions, ±5 degrees.

DESCRIPTION:
1. Crew actions.
    a. The P* will focus primarily outside the aircraft to clear the aircraft throughout the approach and landing. The P* will announce his intent to perform a roll-on landing, the intended point of landing, and any deviation from the approach.
    b. The P/FE will verify the brakes are released before starting the approach.
    c. The P and NCMs will confirm the suitability of the landing area as requested and will assist the P* in clearing the aircraft to warn of any traffic or obstacles.
2. Procedures.
    a. Before starting the approach and touchdown.

    (1) The P/FE will verify the brakes are released. When the desired approach angle is intercepted, the P* will lower the collective as required to establish the descent and adjust as necessary to maintain a constant angle of approach.

    (2) The P* will maintain entry airspeed and aircraft in trim, until reaching approximately 100 feet AGL or a point from which the obstacles can be cleared. The P* will then assume a decelerating attitude (approximately 5 to 10 degrees, nose high) to effect a touchdown on the main landing gear.

    (3) The NCM will inform the P* when the rear of the aircraft is clear of all obstacles in the flight path.

    (4) The P* will slip the aircraft during the deceleration to achieve runway alignment upon touchdown.

    (5) The P* will maintain the desired angle of descent with the collective. Prior to touchdown, the P* will adjust the collective control to affect a smooth touchdown on the main landing gear.

Note. For training, establish entry airspeed 70 KIAS, ±10 KIAS.

    b. After landing.

    (1) The P* will maintain the landing attitude with the cyclic and collective control (not to exceed 10 degrees nose high) until forward speed is sufficiently slowed or stopped. The P* will smoothly lower the collective until the nose gear contacts the ground.

    (2) The P* will then neutralize the flight controls and apply brakes as necessary to stop forward movement.

Note. Do not use aerodynamic braking to slow the aircraft down once the nose gear is in contact with the ground.

```
┌─────────────────────────────────────────────────────────┐
│                        WARNING                          │
│                                                         │
│   Do not allow the tail bumper to contact the ground    │
│   during this procedure.                                │
└─────────────────────────────────────────────────────────┘
```

```
┌─────────────────────────────────────────────────────────┐
│                        CAUTION                          │
│                                                         │
│   Ensure correct fuselage center-line lance alignment   │
│   prior to allowing nose gear to contact the surface.   │
│   Faulty alignment can cause nose gear "wobble" which   │
│   may result in damage to the nose gear.  Should wobble │
│   occur, decrease collective sufficiently to place      │
│   weight on the nose gear, apply brake, align the       │
│   fuselage with ground track.                           │
└─────────────────────────────────────────────────────────┘
```

ROUGH/UNPREPARED SURFACE CONSIDERATIONS: Closely monitor touchdown speed when landing to a rough or unprepared surface. Consistent with the situation and aircraft capabilities, a more pronounced deceleration before touchdown coupled with stronger aerodynamic braking after touchdown may be appropriate.

Note. The wheel brakes may be less effective. If the surface is soft, exercise care when lowering the collective until the aircraft comes to a complete stop.

NIGHT OR NIGHT VISION GOGGLE CONSIDERATIONS: Altitude, apparent ground speed, and rate of closure are difficult to estimate at night. After establishing the descent, the P* should reduce airspeed to approximately 70 KIAS and maintain airspeed until the apparent ground speed and rate of closure appear to be increasing. The rate of descent at night during the final 100 feet should be slightly slower than during the day to avoid abrupt attitude changes at low altitudes. The P* should progressively decrease the rate of descent and forward speed until he or she terminates the maneuver.

TRAINING AND EVALUATION REQUIRMENTS:

1. Training will be conducted in the aircraft or a Mi-17 FS.
2. Evaluation will be conducted in the aircraft.

REFERENCES: Appropriate common references.

TASK 1068

Perform Go-Around

CONDITIONS: In a Mi-17 helicopter or a Mi-17 FS.

STANDARDS: Appropriate common standards and the following additions/modifications:

1. Determine when a go-around is required.
2. Immediately apply climb power (not to exceed aircraft limits).
3. Accelerate to climb airspeed, ±10 knots.
4. Maintain aircraft in trim.
5. Maintain appropriate ground track.

DESCRIPTION:

1. Crew actions.
   a. The P* will announce the intent to perform a go-around and will remain primarily focused outside to avoid obstacles.
   b. The P and the NCMs will assist in clearing the aircraft and provide adequate warning of obstacles. The P/FE will also monitor systems instruments to ensure aircraft limits are not exceeded.
   c. The P or NCMs may call for a go-around if they detect an unsafe landing area. The P* will acknowledge and initiate a go-around.
2. Procedures.
   a. When it becomes doubtful that a safe landing can be accomplished, announce "go-around." Immediately increase power and simultaneously adjust pitch attitude to stop the descent and start a climb to clear any obstacles.
   b. Maintain aircraft in trim and accelerate to the appropriate climb speed for conditions.
   c. Maintain the appropriate ground track.

   Note. The decision to go-around may be made at any time, but in limited power situations should be determined before descending below the barriers or decelerating below ETL.

NIGHT OR NIGHT VISION GOGGLE CONSIDERATIONS: A go-around should be initiated if visual contact with the landing area is lost.

SNOW/SAND/DUST CONSIDERATIONS: If, during the approach, visual reference with the landing area or obstacles is lost, initiate a go-around immediately. Be prepared to transition to instruments and perform an instrument takeoff. Once VMC is regained, continue with the go-around.

MOUNTAINOUS AREA CONSIDERATIONS: If, at any time during an approach, insufficient power is available or turbulent conditions or wind shift create an unsafe condition, perform a go-round immediately. Perform one of the following:

1. Where escape routes exist, turn the aircraft away from the terrain, apply forward cyclic, and lower the collective, if possible. Accelerate the aircraft to an appropriate airspeed for conditions and complete the go-around.
2. Where escape routes do not exist, adjust aircraft for maximum rate of climb to ensure obstacle clearance. Upon clearing obstacles, accelerate aircraft to an appropriate airspeed for conditions and complete the go-around.

TRAINING AND EVALUATION REQUIREMENTS:

1. Training will be conducted in the aircraft or a Mi-17 FS.
2. Evaluation will be conducted in the aircraft.

REFERENCES: Appropriate common references.

TASK 1070

Respond to Emergencies

CONDITIONS: In a Mi-17 helicopter, a Mi-17 FS, or academically and given a specific emergency condition or the indications of a specific malfunction.

> *Note.* Only qualified and current IPs/SPs may simulate emergency procedures when at one set of flight controls in the aircraft only.

STANDARDS: Appropriate common standards and the following additions/modifications:

1. RCM.
    a. Recognize, announce, and analyze indication of an emergency. Perform or describe the immediate action emergency checks IAW the PM-NSRWA manuals or flight manual.
    b. Perform appropriate emergency procedure.
    c. Confirm suitability of the landing area if required.

2. NCM.
    a. Recognize, announce, and analyze indication of an emergency. Perform or describe the immediate action emergency checks IAW the PM-NSRWA manuals or flight manual.
    b. Perform appropriate emergency procedure.
    c. Prepare the aircraft and passengers for an emergency landing.
    d. Assist in confirming the suitability of the landing area if required.
    e. Assist in evacuating passengers to designated assembly area.

DESCRIPTION:

1. Crew actions. A crewmember detecting an emergency will immediately announce the emergency to the other crewmembers.
    a. The crew will perform the steps as appropriate IAW the flight manual/CL and initiate the appropriate type of landing if required.

    (1) During VMC, the P* will focus primarily outside the aircraft to maintain aircraft control and obstacle clearance.

    (2) During IMC, the P* will remain focused inside the aircraft on the flight instruments to maintain aircraft control. If time permits, RCMs will also lock shoulder harnesses, make a mayday call, and tune the transponder to emergency as required.

    b. If time permits, the P/FE will verify all emergency checks with the flight manual/CL. The P/FE will request appropriate emergency assistance.
    c. The cabin NCM will prepare the passengers for an emergency landing, ensuring passengers' seatbelts are fastened and cargo is secured.

    (1) During descent, the NCMs will assist in clearing the aircraft.

    (2) After landing, the NCMs will assist in evacuating the passengers to the designated assembly area. If normal exits cannot be used, the NCMs will use the nearest emergency exit to expedite the evacuation.

    (3) After accounting for all crewmembers and passengers, the NCMs will assist the other crewmembers in any follow-on action (fire fighting, first aid, emergency signaling, or survival equipment).

2. Procedures.
    a. Analyze the emergency situation (for example, aircraft response and caution light indications).
    b. Determine the malfunction and select the appropriate emergency procedures IAW the flight manual/CL.

NIGHT OR NIGHT VISION GOGGLE CONSIDERATIONS: Take special precautions to identify the correct switches/levers when performing emergency procedures at night or while wearing NVG.

TRAINING AND EVALUATION REQUIREMENTS:

1. Training will be conducted in the aircraft, a Mi-17 FS, or academically.
2. Evaluation will be conducted in the aircraft, a Mi-17 FS, or academically.

REFERENCES: Appropriate common references.

TASK 1074

Respond to Engine Failure at Cruise Flight

CONDITIONS: In a Mi-17 helicopter or a Mi-17FS with an IP at one set of the flight controls.

STANDARDS: Appropriate common standards and the following additions/modifications:

1. Identify the emergency, determine the appropriate corrective action, and perform, from memory, all immediate action procedures IAW the flight manual/CL.

2. Adjust the collective to maintain rotor within limits.

3. Maintain airspeed between maximum and minimum autorotation glide speed.

4. Verify that the emergency procedure has been correctly accomplished IAW the flight manual/CL.

5. Select a suitable landing area.

DESCRIPTION:

1. Crew actions:

   a. The IP initiates the maneuver. If in the left seat the IP may simulate the failure by decreasing the ECL on one engine. If in the right seat the IP may direct the pilot in the left seat to decrease an ECL to simulate an engine failure.

   b. The P* will perform, or direct the P/FE to perform, the immediate action steps (underlined) in the Mi-17 CL.

   c. The P/FE will perform as directed or briefed. The P/FE will monitor cockpit instruments to warn the P* prior to exceeding any aircraft limitations. The P/FE will verify each emergency check with the Mi-17 CL. The P/FE will simulate requesting appropriate emergency assistance (simulated) as required.

   d. The NCM will prepare the passengers for an emergency landing, ensuring passengers' seatbelts are fastened and cargo is secured.

   (1) During the descent, the NCM will assist in clearing the aircraft.

   (2) After landing, the NCM will assist in evacuating the passengers to the designated assembly area. If normal exits cannot be used, the NCM will use the nearest emergency exit to expedite the evacuation.

   (3) After accounting for all crewmembers and passengers, the NCM will assist the other crewmembers in any follow-on action (fire fighting, first aid, emergency signaling, or survival equipment).

2. Procedures. The IP will announce "simulated engine failure." Upon detecting and verifying a "simulated" engine failure, the P* will immediately evaluate and determine if continued flight is possible. He or she will then perform the emergency procedure IAW the Mi-17 CL and advise crewmembers of intentions. Complete a landing as appropriate per direction of the IP.

NIGHT OR NIGHT VISION GOGGLE CONSIDERATIONS: Take special precautions to identify the correct switches/levers when performing emergency procedures at night or while wearing NVG.

TRAINING AND EVALUATION REQUIREMENTS: Training and evaluation of this maneuver may include the simulated introduction of an engine failure by reducing one engine to ground idle.

1. Training will be conducted in the aircraft or a Mi-17 FS.

2. Evaluation will be conducted in the aircraft or a Mi-17 FS.

REFERENCES: Appropriate common references.

TASK 1075

Perform Single-Engine Landing

```
WARNING

Prior to performance of the maneuver, the IP must verify with the
performance planning data that the aircraft can be operated within
single-engine limitations.
```

```
CAUTION

Ensure correct fuselage center-line alignment prior to allowing nose gear to
contact the surface. Faulty alignment can cause nose-gear "wobble" which
may cause damage to the nose gear. Should wobble occur, decrease
collective sufficiently to place weight on the nose gear, apply brake, and
align the fuselage with the ground track.
```

CONDITIONS: In a Mi-17 helicopter or a Mi-17 FS, with an IP, and the before-landing check completed; given entry altitude and airspeed.

STANDARDS: Appropriate common standards plus these additions/modifications:

1. Select suitable landing area.

2. Maintain constant approach angle clear of obstacles to desired point of touchdown.

3. Maintain ground track alignment. Execute a smooth, controlled touchdown at speed appropriate for the conditions, but not exceeding 27 Knots ground speed.

4. Touchdown with a maximum of 10 degrees nose high attitude aligned with the landing direction +- 5 degrees.

DESCRIPTION:

1. On downwind leg, the P* will reduce the collective to achieve a power setting to allow single engine operations and adjust cyclic to attain the appropriate airspeed while maintaining altitude. To simulate single engine failure, decrease the ECL on one engine while maintaining proper RPM. Confirm that PTIT and N_G are within limits. At the entry point, establish the desired approach by reducing collective and adjusting cyclic as necessary to achieve the proper airspeed. Maintain 60 to 80 KIAS until reaching approximately 100 feet AGL or a point that will clear any obstacles. Assume a decelerating attitude, while continuing to maintain a constant approach angle. Affect a smooth touchdown at or below 27 knots ground speed without exceeding single-engine PTIT or Ng limitation.

2. After touchdown, neutralize the flight controls , maintain ground track with pedals, and apply brakes as necessary.

TRAINING AND EVALUATION REQUIREMENTS:

1. Training will be conducted in the aircraft or a Mi-17 FS.

2. Evaluation will be conducted in the aircraft or a Mi-17 FS.

REFERENCES: Appropriate common references.

TASK 1082

Perform Autorotation

CONDITIONS: In a Mi-17 helicopter or a Mi-17 FS, with an IP and the before-landing check completed and given entry altitude and airspeed.

STANDARDS: Appropriate common standards and the following additions/modifications:
1. Establish entry altitude as directed, ±100 feet.
2. Establish entry airspeed as directed, ±10 KIAS.
3. Establish correct lane/landing ground track.
4. Select the correct entry point.
5. Visually check N_R, engine speed (N_G) steady rate of descent, and aircraft in trim.
6. Ensure throttle is returned to full on (full right) by 500 feet AGL.
7. Ensure the airspeed at 125 feet AGL is not less than 80 KIAS.
8. Execute a deceleration.
9. Execute a termination as directed by the IP.

DESCRIPTION:
1. Crew actions.
   a.   The P* will remain focused primarily outside the aircraft throughout the maneuver and announce when the maneuver is initiated. The P* will announce the intended point of termination. Prior to initiating the autorotation, the P* will direct the P/FE to monitor N_R, N_G, aircraft trim, altitude/radar altitude, and airspeed. The P* will announce initiation of the autorotation and any deviation during the autorotation.
   b.   The P/FE will announce adequate warning for corrective action if limits for N_R, N_G, steady rate of descent and aircraft trim, or airspeed may be exceeded.
   c.   The IP will announce "power recovery," "roll on landing," or "terminate with power."
2. Procedures.
   a.   When the P* reaches the correct entry point, he or she will smoothly lower the collective to the full down position. The P* will apply pedals required to and obtain a stabilized hover, apply cyclic as required to assume an 80-knot attitude and initiate a turn as required (bank angle not to exceed 20 degrees). When aligned with the runway, the P* will reduce the throttle to the ground idle position (full left). The P* will ensure that N_R is in the normal range.
   b.   During the descent, the P* will closely monitor N_R for an over-speeding condition and adjust the collective as appropriate. The P* will maintain 80 KIAS and the aircraft in trim during the descent. By 500 feet AGL the IP will announce and the P* will acknowledge the type of termination. Prior to passing through 500 feet AGL, the P* will smoothly turn the throttle to flight idle (full right). If any of the above conditions are not met, the P*/IP will correct the condition(s) and execute a go-around. Between 100 and 125 feet AGL, the P* will apply aft cyclic as necessary to assume a deceleration attitude.
   c.   Power recovery. Upon receiving the command "power recovery" the P* will increase the throttle to flight idle (full right) and adjust the collective as necessary while simultaneously maintaining trim with the pedals. The P* will apply sufficient collective to establish a normal climb prior to reaching 200 feet AGL.
   d.   Roll on landing. Upon receiving the command "roll on landing," the P* will increase the throttle to flight idle (full right), adjust the collective as necessary to maintain rotor RPM and trim with the pedals. At 100 to 120 feet AGL, decelerate the aircraft by adding 15 to 20 degrees nose up pitch to arrive at 30 to 40 feet AGL/30 KIAS. Put the aircraft into a landing attitude, 5 to 10 degrees nose high prior to touchdown. Complete a roll on landing IAW Task 1064.
   e.   Terminate with power. Upon receiving the command "terminate with power," the P* will increase the throttle to flight idle (full right) and adjust the collective as necessary, trim the aircraft with the pedals, and maintain autorotation. During the final portion of the maneuver, the P* will apply sufficient power to arrest the descent and obtain a stabilized hover. Ground speed at this point should be below 27 knots.

NIGHT OR NIGHT VISION GOGGLE CONSIDERATIONS: Training in this maneuver at night or when crewmembers are wearing NVG is only to be conducted during IP training or method of instruction training.

TRAINING AND EVALUATION REQUIREMENTS:

1. Training will be conducted in the aircraft or a Mi-17 FS.
2. Evaluation will be conducted in the aircraft.

REFERENCES: Appropriate common references.

TASK 1094

Perform Flight with Auto-Pilot System Off

CONDITIONS: In a Mi-17 helicopter or a Mi-17 FS, with the auto-pilot system off and under VMC.

STANDARDS: Appropriate common standards and the following additions/modifications:
1. RCM. Maintain task standards for the maneuver being performed.
2. FE/NCM. Verify crew, passengers, cargo, and mission equipment are properly secured.
3. Maintain flight with aircraft in trim ±1 ball width.

DESCRIPTION:
1. Crew actions. The P* or P will announce to the other crewmembers when he or she detects an auto-pilot system malfunction. The P* will react positively and smoothly to divergent movements, enter all maneuvers slowly, and avoid over controlling the aircraft. During VMC, the P* will focus primarily outside the aircraft to maintain aircraft control and obstacle clearance. If necessary, the P* will direct the P/FE to disengage the auto-pilot system.

2. Procedures. The P* will smoothly coordinate control movements to maintain the aircraft in trim. The P* will monitor the turn-and-slip indicator for indications of divergent movements. The P* will smoothly and positively react to any divergent movements of the aircraft. The FE/NCM will check that all passengers are wearing their seatbelts and all cargo and mission equipment is secured.

Note. Any maneuver in this ATM may be conducted with the auto-pilot system off **except for** external load hook-up and combat maneuvering flight. The standards for these maneuvers are the same as with the auto-pilot system on.

NIGHT OR NIGHT VISION GOGGLES CONSIDERATIONS: To aid in preventing spatial disorientation, **do not** make large or abrupt attitude changes.

TRAINING AND EVALUATION REQUIREMENTS:
1. Training may be conducted in the aircraft or a Mi-17 FS.
2. Evaluation will be conducted in the aircraft.

REFERENCES: Appropriate common references.

TASK 1114

Perform Rolling Takeoff

CONDITIONS: In a Mi-17 helicopter or a Mi-17 FS, given a suitable takeoff area.

STANDARDS: Appropriate common standards and the following additions/modifications:

1. Prior to the liftoff.
    a. Establish and maintain power, as necessary.
    b. Maintain alignment with takeoff direction, ±5 degrees.
    c. Accelerate to desired liftoff speed not to exceed 27 knots ground speed.
2. After the liftoff.
    a. Adjust power, as required, not to exceed aircraft limits.
    b. Maintain ground track alignment with the takeoff direction with minimum drift.
    c. Maintain directed rate of climb airspeed, ±5 KIAS.
    d. Maintain aircraft in trim above 50 feet.

DESCRIPTION:

1. Crew actions.
    a. The P* will remain focused primarily outside the aircraft during the maneuver. The P* will announce when he or she initiates the maneuver and the intent to abort or alter takeoff.
    b. The P and NCM will announce when ready for takeoff and remain focused primarily outside the aircraft to assist in clearing and provide adequate warning of obstacles. The FE will announce when ready for takeoff and remain focused primarily inside the aircraft. The P will announce when his or her attention is focused inside the cockpit. The FE will monitor power requirements, and ground speed, and will advise the P* when power limits are being approached.
2. Procedures.
    a. A rolling takeoff is used when hover power for takeoff is marginal or insufficient and a takeoff must be made. The concept is to use rotor system power to accelerate the aircraft to a more efficient speed while not having to produce lift sufficient for flight.
    b. Verify the takeoff surface is suitable for the maneuver and select ground reference points. Neutralize the cyclic and raise the collective to establish the aircraft light on the wheels. Use the pedals to maintain heading. Coordinate forward cyclic and raise the collective as necessary to accelerate the aircraft. Maintain heading with pedals and accelerate, not to exceed 27 knots ground speed. Upon reaching lift-off speed, adjust power and cyclic as necessary to allow the aircraft to become airborne. After lift-off, trim the aircraft as soon as possible. Establish and maintain desired rate of climb airspeed until the aircraft is clear of obstacles.

Note. For training, obtain hover blade pitch setting prior to conducting simulated max gross weight situations requiring a rolling take off. Use 3 degree blade pitch to initiate roll and 1 degree blade pitch below hover as maximum available.

Note. Pilot technique, winds, and type of runway surface will affect the distance needed to perform this maneuver.

Note. Aircraft tends to come off the surface nose-gear last, which may result in excessive nose low condition. As collective is increased to lift off, adequate aft right cyclic will bring all gear off simultaneously.

NIGHT OR NIGHT VISION GOGGLE CONSIDERATIONS:

1. If sufficient illumination or NVD resolution exists to view obstacles, accomplish takeoff in the same way as a rolling takeoff during the day. Visual obstacles such as shadows should be treated as physical obstacles. If sufficient illumination or NVD resolution does not exist, a rolling takeoff should not be performed.

2. Reduced visual references during takeoff and throughout ascent at night may make it difficult to maintain the desired ground track. Knowledge of surface wind direction and velocity will assist in establishing the crab angle required to maintain the desired ground track.

TRAINING AND EVALUATION REQUIREMENTS:

1. Training will be conducted in the aircraft or a Mi-17 FS.
2. Evaluation will be conducted in the aircraft.

REFERENCES: Appropriate common references.

TASK 1155

Negotiate Wire Obstacles

CONDITIONS: In a Mi-17 helicopter, a Mi-17 FS, or academically.

STANDARDS: Appropriate common standards and the following additions/modifications:

1. Locate and accurately estimate the height of wires.
2. Determine the best method to negotiate the wire obstacle.
3. Safely negotiate the wire obstacle, minimizing the time unmasked.

DESCRIPTION:

1. Crew actions.
    a. The P* will remain focused primarily outside the aircraft.
    b. The P and NCMs will announce adequate warning to avoid hazards, wires, and poles or supporting structures. They also will announce when the aircraft is clear and when their attention is focused inside the aircraft.
2. Procedures.
    a. Announce when wires are seen. Confirm the location of wire obstacles with other crewmembers.
    b. Discuss the characteristics of wires and accurately estimate the amount of available clearance between the wires and the ground to determine the preferred method of crossing the wires. Locate guy wires and supporting poles.
    c. Announce the method of negotiating the wires and when the maneuver is initiated. Before crossing the wires, identify the highest wire. Cross near a pole to aid in visual perception and minimize the time the aircraft is unmasked. When under-flying wires, maintain a minimum clearance of hover height plus 30 feet and a ground speed no greater than that of a brisk walk. Ensure lateral clearance from guy wires and poles.

Note. The crew must maintain proper scanning techniques to ensure obstacle avoidance and aircraft clearance.

NIGHT OR NIGHT VISION GOGGLE CONSIDERATIONS: Wires are difficult to detect with NVG.

TRAINING AND EVALUATION REQUIRMENTS:

1. Training will be conducted in the aircraft or a Mi-17 FS.
2. Evaluation will be conducted in the aircraft or a Mi-17 FS.

REFERENCES: Appropriate common references.

TASK 1162

Perform Emergency Egress

WARNING

Removing an injured crewmember or passenger may increase the severity of the injuries. Analyze the risk of additional injury versus the risk of leaving the crewmember or passenger in the aircraft until assistance arrives.

CONDITIONS: In a Mi-17 helicopter or academically.

STANDARDS: Appropriate common standards and the following additions/modifications:

1. Perform or describe the use of emergency exits IAW the flight manual.
2. Perform or describe the emergency egress of pilot, NCM, or passenger from his seat.
3. Perform or describe the emergency engine shutdown IAW the flight manual.
4. Marshall passengers to designated assembly area.
5. Perform or describe duties as briefed in the crew mission briefing.

DESCRIPTION:

1. Crew actions.
   a. The PC will direct an emergency evacuation. The PC will determine if the evacuation will be accomplished before the rotor blades have stopped. (If the PC is incapacitated, the next ranking RCM/NCM will perform this function.) The PC will also determine and announce if an emergency engine shutdown will be performed.
   b. The RCMs and FE will egress their respective positions and assist with passenger egress.
   c. The NCMs will direct passenger egress.
   d. All crewmembers will perform duties as briefed during the crew briefing and assist with the egress of incapacitated crewmembers and passengers, if required.
2. Procedures.
   a. If an emergency egress occurs, use the cabin/cockpit doors. If they are jammed, use the emergency release. If the emergency release does not work, break out the windows with the crash axe, boot, or other suitable object. Once out, guide yourself and passengers to clear the aircraft in a safe direction and meet at the assembly point.
   b. Account for all personnel.
   c. Perform the emergency egress of a pilot from his or her seat IAW the flight manual.
   d. Perform the emergency engine shutdown procedures IAW the flight manual.

OVERWATER CONSIDERATIONS: Secure a handhold within the cockpit to maintain orientation, employ the underwater breathing device (if equipped), and wait for the cockpit and cabin area to fill with water. Once the aircraft is full of water, use the cabin/cockpit doors. If they are jammed, use the emergency release. If the emergency release does not work, break out the windows with the crash axe, boot, or other suitable object. Swim clear of the aircraft; **do not** activate the life preserver until clear of the aircraft and on the surface.

WARNING

When an evacuation is performed with the rotor blades turning, **BEWARE** of making contact with the rotor blades. If egress must be made from an aircraft that has gone into the water, <u>do not</u> exit until rotor blades have stopped.

TRAINING AND EVALUATION REQUIREMENTS:

1. Training will be conducted in the aircraft or in a classroom.
2. Evaluation will be conducted in the aircraft or in a classroom.

REFERENCES: Appropriate common references.

TASK 1166

Perform Instrument Maneuvers

CONDITIONS: In a Mi-17 helicopter or a Mi-17 FS, while wearing a hood or during IFR.

STANDARDS: Appropriate common standards and the following additions/modifications:
1. Straight and level.
    a. Altitude maintained, ±100 feet.
    b. Heading maintained, ±10 degrees.
    c. Airspeed maintained, ±10 KIAS.
    d. Aircraft maintained in trim.
2. Climb/Descent.
    a. Clearance acknowledged-climb/descent initiated as soon as practical.
    b. Obtained/maintained airspeed, ±10 KIAS.
    c. Power adjusted to climb/descend as rapidly as practical to ±1000 feet of assigned altitude, then 500 FPM (±100 FPM).
    d. Level off at assigned altitude, ±100 feet.
    e. Heading maintained (unless turning), ±10 degrees.
    f. Aircraft maintained in trim.
3. Turns.
    a. Standard rate turn-two needle width deflection of the turn needle obtained/maintained (4-minute turn and slip) and turns in the correct direction.
    b. Half standard rate turn-one needle width deflection of the turn needle obtained/maintained.
    c. Desired airspeed maintained, ±10 KIAS.
    d. Timed turns-recovery to straight and level flight based on time rather than slaved gyro-compass indications.
    e. Assigned altitude maintained (unless climbing or descending), ±100 feet.
    f. Rollout made on assigned/desired heading, ±10 degrees.
    g. Aircraft maintained in trim.
4. Radio navigation.
    a. Tune and identify appropriate navigation aids (NAVAIDs).
    b. Determine, intercept, and maintain the desired course IAW FM 3-04.240 and FAR part 91.
    c. Identify station passage.
    d. For area navigation and direct routing, ensure flight route meets minimum en route altitude requirements.

DESCRIPTION: For a detailed description refer to FM 3-04.240.

TRAINING AND EVALUATION REQUIREMENTS:
1. Training will be conducted in the aircraft or a Mi-17 FS.
2. Evaluation will be conducted in the aircraft or a Mi-17 FS.

REFERENCES: Appropriate common references.

TASK 1170

Perform Instrument Takeoff

CONDITIONS: In a Mi-17 helicopter or a Mi-17 FS, under IFR, with reference to instruments only, with hover power check and before-takeoff checks completed, and aircraft cleared.

STANDARDS: Appropriate common standards and the following additions/modifications:

1. Correctly determine takeoff power.
2. Maintain power as required.
3. Maintain accelerative climb attitude, ±1 bar width (not to exceed 10 degrees nose low), until climb airspeed is attained.
4. Maintain takeoff heading, ±10 degrees.
5. Maintain the aircraft in trim after 40 KIAS.
6. Maintain an appropriate rate of climb, ±200 FPM.
7. Maintain desired climb airspeed, ±10 KIAS.

DESCRIPTION:

1. Crew actions.
    a. The P* will focus primarily outside the aircraft during the VMC portion of the maneuver. The P* will announce when they initiate the maneuver and any intentions to alter or abort takeoff.
    b. The P/NCMs will announce when ready for takeoff and will focus primarily outside the aircraft to assist in clearing during the VMC portion of the maneuver and to provide adequate warning of obstacles.
2. Procedures.
    a. From the ground.

    (1) Align the aircraft with the desired takeoff heading. Ensure the attitude indicator is set for takeoff.

    (2) Initiate takeoff by increasing the collective smoothly and steadily, while maintaining a level attitude, until instrument-takeoff power is reached. When instrument-takeoff power is established and the altimeter and VSI show a positive climb, adjust pitch attitude below the horizon as required for the initial acceleration (not to exceed 10-degrees nose low).

    (3) Visually maintain runway clearance and alignment on takeoff and transition to flight instruments before entering IFR. At approximately 40 KIAS, the P* will check the turn-and-slip indicator to ensure the aircraft is in trim.

    (4) Maintain the heading/course required by the departure procedure or ATC instructions. When the desired climb airspeed is reached, adjust cyclic to maintain airspeed and adjust the collective to maintain the desired climb rate.

    b. From a hover.

    (1) The P* will align the aircraft with the desired takeoff heading at the appropriate hover height. The P* will check the attitude indicator for the appropriate attitude.

    (2) The P* will initiate the takeoff by increasing the collective smoothly and steadily, while maintaining a level attitude, until instrument-takeoff power is reached.

    (3) When the altimeter and VSI show a positive rate of climb, the P* will continue as in a takeoff from the ground.

Note. Performing this maneuver in certain environments may require hover OGE power. Evaluate each situation for power required versus power available.

Note. When the crew is operating under IFR, the NCM will take a position on the P*'s side of the aircraft for obstacle clearance and airspace surveillance.

TRAINING AND EVALUATION REQUIREMENTS:

1. Training may be conducted in the aircraft or a Mi-17 FS.
2. Evaluation may be conducted in the aircraft or a Mi-17 FS.

REFERENCES: Appropriate common references.

TASK 1174

Perform Holding Procedures

CONDITIONS: In a Mi-17 helicopter or in a Mi-17 FS under IFR and given holding instructions and appropriate DOD FLIPs.

STANDARDS: Appropriate common standards and the following additions/modifications:

1. Tune and identify the appropriate NAVAIDs.
2. Enter the holding pattern.
3. Time and track holding pattern legs.
4. Send the appropriate report to ATC IAW DOD FLIPs.

DESCRIPTION:

1. Crew actions.

   a. Prior to arrival at the holding fix, the PC will analyze the holding instructions and determine the holding pattern and proper entry procedures. The PC will brief the other crewmembers on the proposed entry, outbound heading, and inbound course. (The PC may delegate this task to another RCM.)

   b. The P will select radio frequencies and monitor radios. The P will announce ATC information not monitored by the P*. The P also will compute outbound times and headings to adjust for wind and direct the P* to adjust the pattern as necessary.

   c. The P* will fly headings and altitudes and will adjust inbound and outbound times as directed by ATC or the P. The P* will announce any deviation as well as ATC information not monitored by the P.

   d. During IFR, the RCMs and the FE will focus primarily inside the aircraft. The cabin NCMs will provide adequate warning of traffic or obstacles. They will announce when their attention is focused inside the aircraft and again when attention is reestablished outside.

2. Procedures. Upon arrival at the holding fix, turn (if required) to the predetermined outbound heading or track and check the inbound course. Maintain the outbound heading or track as published or as directed by ATC. After the appropriate time outbound, turn to the inbound heading and apply normal tracking procedures to maintain the inbound course. The P will note the time required to fly the inbound leg and adjust outbound course and time if necessary. When holding at a NAVAID, the P will begin timing the outbound leg when abeam the station. When holding at an intersection, the P will begin timing the outbound leg upon establishing the outbound heading.

TRAINING AND EVALUATION REQUIREMENTS:

1. Training may be conducted in the aircraft or a Mi-17 FS.
2. Evaluation may be conducted in the aircraft or a Mi-17 FS.

REFERENCES: Appropriate common references.

TASK 1176

Perform Nonprecision Approach

CONDITIONS: In a Mi-17 helicopter or a Mi-17 FS, under IFR, with reference to instruments only, given approach information and appropriate DOD FLIP approach clearance, with the before-landing checks complete.

STANDARDS: Appropriate common standards and the following additions/modifications:

1. Execute the approach IAW AR 95-1, FM 3-04.240, AIM, and the DOD FLIP.

2. Maintain very high frequency omni-directional range or GPS course centerline, ±5 degrees.

3. Maintain localizer course centerline, ±2.5 degrees.

4. During airport surveillance radar approaches, make immediate heading and altitude changes issued by ATC and maintain heading, ±5 degrees.

5. Comply with descent minimums prescribed for the approach.

6. Perform the correct missed approach procedure IAW DOD FLIP or ATC instructions upon reaching the missed approach point (MAP), if landing cannot be accomplished IAW AR 95-1.

DESCRIPTION:

1. Crew actions.

   a. Each RCM will review and confirm the specific approach to be flown before initiating the procedure. The crew will confirm the correct NAVAID/communication frequencies; the horizontal situation indicator (HSI) is set as required.

   b. The P* will focus primarily inside the aircraft on the instruments and perform the approach. The P* will follow the heading/course, altitude, and missed approach directives issued by the P/ATC. The P* will announce any deviation not directed by ATC or the P and will acknowledge all navigation directives given by the P.

   c. The P will call out the approach procedure to the P* and will advise the P* of any unannounced deviations. The P will monitor outside for the landing environment; announce when he or she makes visual contact suitable to complete the landing IAW AR 95-1; and, if directed by the P*, take the controls to complete the landing.

   d. The P will announce if he or she does not make visual contact by the MAP and call out the missed approach procedures. During IFR, the P and NCMs will focus primarily outside the aircraft to provide adequate warning of traffic or obstacles. The FE will assume the P* scan sector.

2. Procedures. Perform the desired approach procedures IAW AR 95-1, the appropriate DOD FLIP, FM 3-04.240, and the flight manual.

Note. GPS IFR navigation equipment must be certified by the FAA or host country regulations prior to IFR use. However, they should consider and plan for its use as an emergency backup system only.

TRAINING AND EVALUATION REQUIREMENTS:

1. Training may be conducted in the aircraft or a Mi-17 FS.

2. Evaluation may be conducted in the aircraft or a Mi-17 FS.

REFERENCES: Appropriate common references.

TASK 1178

Perform Precision Approach

CONDITIONS: In a Mi-17 helicopter or a Mi-17 FS, under IFR, with reference to instruments only, given approach information and the appropriate DOD FLIP approach clearance, and the before-landing checks complete.

STANDARDS: Appropriate common standards and the following additions/modifications:

1. Execute the approach IAW AR 95-1, FM 3-04.240, AIM, and the DOD FLIP.

2. For an instrument landing system approach, maintain the localizer centerline, ±2.5 degrees, and the glide slope indicator within full scale deflection.

3. For a precision approach radar (PAR) approach, make immediate heading and altitude changes issued by ATC and maintain heading, ±5 degrees; for final approach, maintain glide slope as directed by ATC.

4. Comply with the decision height (DH) prescribed for the approach.

5. Perform the correct MAP IAW the appropriate DOD FLIP or ATC instruction upon reaching the DH if landing cannot be accomplished IAW AR 95-1.

DESCRIPTION:

1. Crew actions.
   a. Each RCM will review and confirm the specific approach to be flown before initiating the procedure. The crew will confirm the correct NAVAID, communication frequencies, and HSI are set as required.
   b. The P* will focus primarily inside the aircraft on the instruments and perform the approach. The P* will follow the heading/course, altitude, and missed approach directives issued by the P/ATC. The P* will announce deviations not directed by ATC or the P and will acknowledge all navigation directives given by the P.
   c. The P will call out the approach procedure to the P* and will advise the P* of unannounced deviations.
   d. The P will monitor outside for the landing environment; announce when he or she makes visual contact suitable to complete the landing IAW AR 95-1; and, if directed by the P*, take the controls to complete the landing. The P will announce if he or she does not make visual contact by the MAP and call out the missed approach procedures.
   e. During IFR, the P and NCMs will focus primarily outside the aircraft to provide adequate warning of traffic or obstacles. The cabin NCM will take a position on the P*'s side of the aircraft.

2. Procedures. Perform the desired approach procedures IAW AR 95-1, the appropriate DOD FLIP, FM 3-04.240, and the flight manual.

> *Note.* GPS IFR navigation must be certified by the FAA or host country regulations prior to GPS IFR navigation; however, they should consider and plan for its use as an emergency backup system only.

TRAINING AND EVALUATION REQUIREMENTS:

1. Training may be conducted in the aircraft or a Mi-17 FS.

2. Evaluation may be conducted in the aircraft or a Mi-17 FS.

REFERENCES: Appropriate common references.

TASK 1180

Perform Emergency Global Positioning System Recovery Procedure

CONDITIONS: In a Mi-17 helicopter or a Mi-17 FS under VMC or IFR, given an approved emergency GPS recovery procedure.

STANDARDS: Appropriate common standards and the following additions/modifications:

1. Enter or confirm the appropriate waypoints (initial approach fix [IAF], intermediate approach fix [IF], final approach fix [FAF], MAP) into the navigation system.

2. Execute the procedure IAW an approved recovery procedure.

3. Maintain a briefed airspeed not to exceed 90 KIAS, appropriate for the conditions, during all segments of the approach.

4. Maintain the prescribed course, ±5 degrees.

5. Comply with the descent minimums prescribed for the procedure.

6. Arrive at the minimum descent altitude prior to reaching the MAP.

7. Execute a missed approach upon reaching the MAP if a safe landing cannot be done.

8. During the missed approach, immediately establish a climb using an appropriate rate of climb airspeed (until established at the minimum safe altitude [MSA]).

DESCRIPTION:

1. Before the flight, the crew should review the recovery procedure in conjunction with the map to familiarize themselves with the procedure and with local terrain and obstructions in the vicinity of the procedure. The PC performs a thorough map reconnaissance to determine the highest obstruction in the area of operations.

2. Prior to initiating the procedure, the P* must climb to the prescribed MSA, proceed toward the IAF, and make the appropriate radio calls. During the procedure, the P* will focus primarily inside the aircraft on the instruments. The P* will adjust the aircraft ground track to cross the IAF, IF, and then the FAF on the prescribed course. When over the FAF, the P* begins the final descent as appropriate.

3. The P/NCMs remain primarily focused outside the aircraft to provide adequate warning for avoiding obstacles/hazards and will announce when his or her attention is focused inside the cockpit. The P and FE will monitor the aircraft instruments during the procedure, and the P will tune the communication and navigation radios and transponder as required. The P will be prepared to call out the procedure to the P*, if asked, and be in a position to assume control of the aircraft and land the aircraft if VMC is encountered.

4. The cabin NCM will position himself on the P*'s side of the aircraft for obstruction clearance and airspace surveillance. The NCMs will alert the crew immediately if VMC is encountered.

NIGHT OR NIGHT VISION GOGGLE CONSIDERATIONS: The P should be in a position to assume control of the aircraft when a landing environment can be determined visually (aided/unaided). During night unaided flight, consider using the landing light to identify the landing area.

TRAINING CONSIDERATIONS: This task will only be performed under VMC.

Note. The IAF, IF, FAF, and MAP should be programmed into the navigation system as an additional route for the mission.

Note. It is not necessary to hold after a missed approach. The PC may elect to return to the IF at the MSA and attempt to complete the approach after coordinating with ATC or with other aircraft using the approach procedure.

Note. IIMC multi-aircraft operations must be thoroughly briefed in the mission brief at a minimum on the following topics: individual aircraft holding altitudes/separation, when individual aircraft are allowed to depart their assigned altitude missed approach procedure with aircraft in the holding pattern; frequencies; and command/control procedures.

TRAINING AND EVALUATION REQUIREMENTS:

1. Training may be conducted in the aircraft or a Mi-17 FS.
2. Evaluation will be conducted in the aircraft or a Mi-17 FS.

REFERENCES: Appropriate common references.

TASK 1182

Perform Unusual Attitude Recovery

CONDITIONS: In a Mi-17 helicopter or a Mi-17 FS, with an IP/IE and under VMC or IFR.

STANDARDS: Appropriate common standards and the following additions/modifications:

1. Analyze aircraft attitude.
2. Without delay, perform recovery procedures in the following sequence:
   a. Attitude-level the wings on the attitude indicator.
   b. Heading-maintain heading; turn only to avoid known obstacles.
   c. Power-adjust power as necessary.
   d. Trim-trim aircraft as necessary.
   e. Airspeed-adjust airspeed as necessary.
3. Clear the aircraft.
4. Recover with a minimum loss of altitude.

DESCRIPTION:

1. Crew actions.
   a. The IP/IE will place the aircraft in unusual attitude and transfer aircraft control to the P. The P will acknowledge the transfer of controls, the unusual attitude, and recover the aircraft as P*.
   b. The P* will remain focused inside the aircraft during this maneuver and will acknowledge the unusual attitude recovery and transfer of aircraft controls.
   c. The P/FE will assist in monitoring the aircraft instruments. The P will call out attitude, power, and trim as necessary.
   d. During VMC, the P and NCMs will focus primarily outside the aircraft to provide adequate warning of traffic or obstacles. They will announce when their attention is focused inside the aircraft and again when attention is reestablished outside.

2. Procedures. To recover from an unusual attitude, correct the pitch and roll attitude, adjust power, and trim the aircraft as required returning to level flight. The displacement of controls used in recoveries may be greater than those for normal flight. Care must be taken in making adjustments as straight-and-level flight is approached. The instruments must be observed closely to avoid over controlling.

Note. NCM tasks may include checking for fire, preparing passengers for an emergency landing, and/or executing any portion of an emergency procedure pertaining to the NCM.

NIGHT OR NIGHT VISION GOGGLE CONSIDERATIONS: Low-level ambient light may induce visual illusions and spatial disorientation. During NVG operations, video noise may contribute to the loss of visual cues.

SNOW/SAND/DUST CONSIDERATION: Obscurants other than weather can induce the loss of visual contact. At low altitudes where these conditions would be encountered, it is extremely important these procedures be initiated to prevent ground contact.

TRAINING AND EVALUATION REQUIREMENTS:

1. Training will be conducted in the aircraft or a Mi-17 FS.
2. Evaluation will be conducted in the aircraft or a Mi-17 FS.

REFERENCES: Appropriate common references.

TASK 1184

Respond to Inadvertent Instrument Meteorological Conditions

CONDITIONS: In a Mi-17 helicopter, a Mi-17 FS under VMC, or academically.

STANDARDS: Appropriate common standards and the following additions/modifications:

1. Announce "IMC," maintain proper aircraft control, and make the transition to instrument flight immediately.
2. Immediately initiate a climb.
3. Continue IMC recovery procedures as follows:
   a. Attitude-level the wings and adjust pitch for desired airspeed while maintaining the aircraft in trim.
   b. Heading-maintain heading; turn only to avoid known obstacles.
   c. Power-maintain climb power until reaching appropriate cruise altitude.
   d. Airspeed-adjust to appropriate climb airspeed.
   e. Altitude-climb to a minimum safe altitude as prescribed by the appropriate DOD FLIP, local regulation, or SOP after establishing aircraft control.
4. Complete the IIMC recovery IAW local regulations and policies.

DESCRIPTION:

1. Crew actions.
   a. The P* will announce "IMC," immediately initiate a climb, and establish aircraft control while transitioning to the instruments.
   b. The P* will immediately announce if he or she becomes disoriented.
   c. The P will announce "IMC," and the P/FE will monitor the cockpit instruments to assist in recovery.
   d. The P* will announce when the aircraft is in a positive climb, the current altitude and altitude climbing to, and the heading.
   e. The P* will adjust the transponder to emergency, adjust the navigational radios as appropriate, and make the appropriate radio calls.
   f. The P performs any other tasks as directed by the P* and will always remain prepared to take the controls should the P* become disoriented.
   g. The cabin NCMs will focus primarily outside the aircraft to provide adequate warning for avoiding terrain or obstacles and will announce if VMC are encountered.
   h. The NCMs will perform any other tasks as directed by the P*/P.
2. Procedures.
   a. The crew should consider establishing climb power and airspeed appropriate for the mission environment to use in the event of encountering IMC. If briefed during the crew briefing, this can help eliminate confusion during the actual emergency.
   b. The most important action when encountering IMC is to immediately begin climbing while establishing aircraft control via the instruments. Once this is accomplished, the transponder should be set to emergency to alert ATC. Tuning navigational radios or making radio calls will be determined by local procedures. The crew should contact ATC on guard and allow ATC to assign an appropriate altitude and heading/course and, if necessary, a frequency. If radio contact cannot be established first, the crew must ensure navigational radios are tuned as quickly as possible to determine the aircraft's position and appropriate course for recovery.

NIGHT OR NIGHT VISION GOGGLE CONSIDERATIONS: When using NVG, it may be possible to see through a thin obscuration, such as fog and drizzle, with little or no degradation. The NVG may be removed (or flipped up) once flight is stabilized.

> *Note.* If IMC are entered with the IR searchlight or landing light on, spatial disorientation may occur. Low-level ambient light may induce visual illusions and spatial disorientation. During NVG operations, video noise may contribute to loss of visual cues.

SNOW/SAND/DUST CONSIDERATIONS: Obscurants other than weather can induce loss of visual contact. At low altitudes where these conditions would be encountered, it is extremely important these procedures be initiated immediately to prevent ground contact.

TACTICAL CONSIDERATIONS: In tactical environments without NAVAIDs, the crew should consider flying a GPS route to a point where an instrument approach (GPS, PAR) is established. The GPS route can be the planned mission route with sufficient terrain/obstacle clearance established in the event of IIMC.

TRAINING AND EVALUATION REQUIREMENTS:
1. Training may be conducted in the aircraft, a Mi-17 FS, or academically.
2. Evaluations will be conducted in the aircraft or a Mi-17 FS.

REFERENCES: Appropriate common references.

TASK 1188

Operate Aircraft Survivability Equipment

CONDITIONS: In a Mi-17 helicopter, a Mi-17 FS equipped with ASE, or academically.

WARNING

ASE systems, when energized, may cause thermal burns or other injuries to personnel too close to an active system. Observe all warnings and cautions. Ensure missile approach warning system (MAWS) or ASE variant safety pin is installed whenever aircraft is in a nonhostile environment or in a position where inadvertent flare/chaff launch may cause injury to personnel and or may cause destruction of equipment or property.

STANDARDS: Appropriate common standards and the following additions/modifications:

1. RCM. Describe the purpose of each installed item of ASE. Perform/describe pre-flight inspection, turn-on, test, operation, emergency procedures, and shutdown of installed ASE. Determine partial failure alternatives. Employ/describe use of installed ASE.

2. NCM. Prepare equipment for operation.

DESCRIPTION:

1. Crew actions.

    a. The PC will ensure crewmembers understand the employment of installed ASE during conduct of the mission. The PC will also ensure all ASE payloads and settings are IAW the mission briefing.

    b. When the crew encounters a radar-directed threat, the P* will remain primarily focused outside to avoid obstacles, perform the required evasive maneuver, reposition the aircraft as necessary to break lock, deploy to cover, and then avoid the threat. The P will dispense chaff prior to performing break-lock evasive maneuvers. The P and NCMs will assist in clearing the aircraft and provide adequate warning of obstacles.

    c. The P will begin dispensing chaff by pressing the chaff dispense button or ensuring the mode switch is in PGRM as required. The P and NCMs will assist in clearing the aircraft and provide adequate warning of obstacles.

    d. When the crew encounters an IR directed threat, the P* will remain primarily focused outside to avoid obstacles, employ evasive maneuvers after defeating the threat with MAWS, deploy to cover, and then avoid the threat. Allow MAWS and variant ASE systems to automatically launch flares. If reliability of equipment is questionable or system has not reacted to observed threat, then P and NCM will launch flares manually.

    e. The NCM will remove and install safety pin(s) IAW the appropriate manual/CL.

2. Procedures.

    a. Perform or describe pre-flight inspection, turn-on, test, operation, emergency procedures, and shutdown of installed ASE equipment.

    b. Evaluate and interpret the ASE visual and aural indications.

    c. Execute mission employment IAW doctrine and determine failure alternatives.

TRAINING AND EVALUATION REQUIREMENTS:

1. Training may be conducted in the aircraft, a Mi-17 FS, or academically.

2. Evaluation will be conducted in the aircraft, a Mi-17 FS, or academically.

REFERENCES: Appropriate common references, computer-based aircraft survivability equipment trainer (CBAT), aircraft survivability equipment trainer (ASET) programs, MAWS, ASE operator manuals, and the unit S-2.

TASK 1190

Perform Hand and Arm Signals

CONDITIONS: Given a list of hand and arm signals from FM 21-60 to identify or perform.

STANDARDS: Appropriate common standards and the following additions/modifications:

1. RCM. Identify, at a minimum, the hand and arm signals required for moving an aircraft left, right, forward, or backward; takeoff; and landing IAW FM 21-60.

2. NCM. Identify and perform, at a minimum, the hand and arm signals required for moving an aircraft left, right, forward, or backward; takeoff; and landing IAW FM 21-60.

DESCRIPTION: Identify or perform the hand and arm signals required to move an aircraft from one point to another.

TRAINING AND EVALUATION REQUIREMENTS:

1. Training will be conducted academically.
2. Evaluation will be conducted academically.

REFERENCES: Appropriate common references and FM 21-60.

TASK 1194

Perform Refueling Operations

```
┌──────────────────────────────────────────────────────┐
│                                                        │
│                       WARNING                          │
│                                                        │
│         Hot refueling operations are prohibited.       │
│                                                        │
└──────────────────────────────────────────────────────┘
```

CONDITIONS: With a Mi-17 helicopter with refueling equipment, or academically.

STANDARDS: Appropriate common standards and the following additions/modifications:

1. Ensure safety procedures are complied with and all individuals are wearing appropriate protective clothing IAW FM 10-67-1, the flight manual/CL, and FM 3-04.113.
2. Ensure the aircraft is refueled IAW FM 10-67-1, the flight manual/CL, FM 3-04.113, and the unit SOP.
3. Enter the appropriate information on DA Form 2408-12 or aircraft documentation as required.

DESCRIPTION: Crew actions, cold refueling. A crewmember will guide the refueling vehicle to the aircraft. Ensure the driver parks the vehicle the proper distance from the aircraft IAW FM 10-67-1. Verify all personnel not involved with the refueling operation are a safe distance away. Ground and refuel the aircraft IAW FM 10-67-1, the flight manual/CL, and the unit SOP. Ensure the tanks are filled to the required level. When the refueling is completed, ensure all caps are secured and remove the ground connection if the aircraft will not remain parked. Make the appropriate entries on DA Form 2408-12 or appropriate approved forms as required.

NIGHT OR NIGHT VISION GOGGLE CONSIDERATIONS: Supplement aircraft lighting at the refueling station by using an unfiltered explosion-proof flashlight to check for leaks and fuel venting.

TRAINING AND EVALUATION REQUIREMENTS:

1. Training will be conducted in the aircraft or academically.
2. Evaluation will be conducted in the aircraft or academically.

REFERENCES: Appropriate common references, FM 10-67-1, and FM 21-60.

TASK 1200

Perform Nonrated Crewmember Duties During Maintenance Test Flight

CONDITIONS: In a Mi-17 helicopter, a Mi-17 FS, or academically and given an Mi-17 MTF manual.

STANDARDS: Appropriate common standards and the following additions/modifications:
1. Perform or describe appropriate maintenance procedures and checks IAW Mi-17 MTF manual.
2. Perform or describe maintenance procedures and checks directed by the MP.
3. Immediately inform the MP of any malfunction or discrepancy detected during the maintenance procedures or checks.

DESCRIPTION:
1. Crew actions.
   a.   If two or more NCMs are performing crew duties during the TF, the MP will ensure they are briefed on their duties and responsibilities.
   b.   NCMs will perform duties and responsibilities as directed by the MP/ME. If any procedure is conducted or the result is not IAW the applicable maintenance or troubleshooting manual, the MP will be notified.
2. Procedures.
   a.   Before and during the TF, the NCMs must constantly monitor all aircraft systems and components. The NCMs will inform the MP of any unusual vibrations, noises, smells, leakage, or component malfunctions. The NCMs will also perform any maintenance procedures and checks required by the MP.
   b. Prior to flight, the NCM will remove any additional panels, covers, and cowlings required by the MP.

TRAINING AND EVALUATION REQUIREMENTS:
1. Training may be conducted in the aircraft, a Mi-17 FS, or academically.
2. Evaluations will be conducted in the aircraft, a Mi-17 FS, or academically.

REFERENCES: Appropriate common references.

TASK 1202

Perform Auxiliary Power Unit Operations

CONDITIONS: In a Mi-17 helicopter or a Mi-17 FS.

STANDARDS: Appropriate common standards and the following additions/modifications:
1. Pre-flight all systems to be operated during APU operations.
2. Operate APU, systems, and equipment IAW the flight manual/CL.
3. Shutdown systems, equipment, and APU IAW the flight manual/CL.
4. Perform or describe appropriate emergency procedures for APU fire IAW the flight manual.
5. Enter appropriate information on DA Form 2408-12, DA Form 2408-13, DA Form 2408-13-1, and as required.

DESCRIPTION:
1. Crew actions.
   a. The crewmember will coordinate with and brief any additional ground support personnel before APU start. Perform pre-flight inspection of the APU. The crewmember ensures the rotor blade tiedowns are removed and the rotor blades are not positioned over the APU exhaust. The crewmember will brief all necessary personnel on procedures to be followed in an emergency. The crewmember will direct assistance from any additional ground support personnel to aid in maintaining the clearance of APU exhaust areas during the APU start sequence and any subsequent ground checks.
   b. Additional ground support personnel should assist the crewmember as directed.
2. Procedures.
   a. Before starting APU, brief the additional ground support personnel as necessary.
   b. Review aircraft logbook for any faults that would prevent operation of the APU, or the APU generator.
   c. Perform pre-flight inspection of the APU and check APU exhaust cover, rotor blade tiedowns, and fluid levels.
   d. Ensure service tank boost pump is on.

NIGHT OR NIGHT VISION GOGGLE CONSIDERATIONS: During night operations, ensure adequate lighting (anti-collision or position lights) is on and the fire guard has a flashlight.

TRAINING AND EVALUATION REQUIREMENTS:
1. Training will be conducted in the aircraft or a Mi-17 FS.
2. Evaluation will be conducted in the aircraft or a Mi-17 FS.

REFERENCES: Appropriate common references.

TASK 1262

Participate in a Crew-Level After Action Review

CONDITIONS: After flight in a Mi-17 helicopter or a Mi-17 FS and given a unit-approved, crew-level (AAR) CL.

STANDARDS: Appropriate common standards and the following additions/modifications:

1. The PC/air mission commander (AMC) will conduct a detailed crew-level AAR using a unit-approved, crew-level AAR CL (table 4-3) after each flight.

2. All crewmembers will actively participate in the review.

Table 4-3. Sample format for a crew-level after action review checklist

1. Restate mission objectives with mission, enemy, terrain and weather, troops and support available, time available, civil considerations (METT-TC).
2. Conduct a review for each mission segment.
a. Restate planned actions/interactions for the segment.
b. What actually happened?
(1) Each crewmember states in own words.
(2) Discuss impacts of crew coordination requirements, aircraft/equipment operation, tactics, commander's intent, and so forth.
c. What was right or wrong about what happened?
(1) Each crewmember states in own words.
(2) Explore causative factors for both favorable and unfavorable events.
(3) Discuss crew coordination strengths and weakness in dealing with each event.
d. What must be done differently the next time?
(1) Each crewmember states in own words.
(2) Identify improvements required in the areas of team relationships, mission planning, workload distribution and prioritization, information exchange, and cross monitoring of performance.
e. What were the lessons learned?
(1) Each crewmember states in own words.
(2) Are changes necessary to the following areas?
(a) Crew coordination techniques.
(b) Flying techniques.
(c) SOP.
(d) Doctrine, ATM, and TMs.
3. Determine the effect of segment actions and interactions on the overall mission.
a. Each crewmember states in own words.
b. Lessons learned.
(1) Individual level.
(2) Crew level.
(3) Unit level.
4. Advise unit operations of significant lessons learned.

DESCRIPTION:

1. Crew actions.

   a. The PC will conduct a crew-level AAR using a unit-approved CL. The PC will actively seek input from all crewmembers. He or she will ensure the results of the review are passed to operations and flight standards.

   b. All crewmembers will actively participate in the review. The intent is to constructively review the mission and apply lessons learned into subsequent missions.

2. Procedures.

   a. Using an AAR CL, participate in a crew-level AAR of the mission. The review should be an open and frank discussion of all aspects of the mission. It should include all mission factors and incorporate all crewmembers.

   b. The results of the review should be passed to operations and flight standards.

TRAINING AND EVALUATION REQUIREMENTS:

1. Training will be conducted academically.

2. Evaluation will be conducted academically.

REFERENCES: Appropriate common references.

TASK 2010

Perform Multi-Aircraft Operations

CONDITIONS: In a Mi-17 helicopter or a Mi-17 FS, with the mission briefing completed.

STANDARDS: Appropriate common standards and the following additions/modifications:

1. RCM.
   a. Participate in a formation flight briefing IAW unit SOP. Table 4-4, page 4-93, lists the minimum items to be briefed.
   b. Maneuver into the flight formation.
   c. Change position in the flight formation when required.
   d. Maintain proper horizontal and vertical separation for the type of formation flight being conducted.
   e. If visual contact is lost, immediately make a radio call to the flight and begin reorientation procedures.
   f. Perform techniques of movement, if required.

2. NCM.
   a. The NCMs assume a position in the helicopter, as briefed, to observe other aircraft in the formation.
   b. Announce if visual contact is lost with other aircraft.

DESCRIPTION:

1. Crew actions.

Note. The most important consideration when an aircraft has lost visual contact with the formation is to announce loss of visual contact and reorientation. Except for enemy contact, all mission requirements are subordinate to this action.

---

   a. The P* will focus primarily outside the aircraft for clearing and tracking other aircraft. The P* will announce any maneuver or movement before execution and inform the P and NCMs if visual contact is lost with other aircraft. If visual contact is lost with other aircraft, complete the following.

   (1) The crew will immediately make a radio call to the flight and begin reorientation procedures (for example, "chalk 3 has loss of visual contact with the formation").

   (2) Lead will announce and maintain heading, altitude, and airspeed until all aircraft have rejoined the flight. The P* will also announce his or her position relative to the next waypoint.

   (3) The aircraft that has lost visual contact with the flight will immediately assume the lead's heading and airspeed, and will maintain vertical separation as briefed.

   (4) If IMC are encountered, execute IIMC breakup as briefed. The P* will ensure that appropriate radio calls are made during IMC breakup.

   b. The P and NCMs will provide adequate warning of traffic or obstacles detected in the flight path and identified on the map. They will inform the P*—

   (1) If visual contact is lost with other aircraft.

   (2) If an enemy is sighted.

   (3) When their attention is focused inside the aircraft.

   (4) When attention is reestablished outside along with the seat position.

   c. The PC will call out direction and altitude in case of IMC breakup. The cabin NCMs will position themselves in the aircraft to observe other aircraft in the formation and to assist in maintaining aircraft separation and obstacle clearance.

2. Procedures.
   a. Perform formation flight IAW the unit SOP and common references in this ATM.
   b. The following procedures will be performed unless otherwise established in unit SOPs.

(1) Takeoff: All helicopters should leave the ground simultaneously. The trailing aircraft must remain at a level altitude or stack up 1 to 10 feet vertically to remain out of the disturbed air of the aircraft in front of them. In the event an aircraft in the flight loses visual contact with the formation, the crew will immediately make a radio call to the formation and the P* will initiate a climb above the briefed cruise altitude and attempt reorientation of the formation.

(2) Cruise: Free cruise formation should be employed when operating at terrain flight altitudes or in a combat environment. This will allow the individual aircraft more flexibility to move within the formation, avoiding terrain, obstacles, and enemy threat. Consideration should be given to door gunner's fields-of-fire to aid in protecting the entire formation. During periods of degraded visibility, crews are more susceptible to losing other aircraft in the formation. Crews should consider flying a close formation to maintain orientation on the flight. In the event an aircraft in the flight loses visual contact with the formation, they will immediately make a radio call to the formation. Lead will announce and maintain heading, altitude, and airspeed. If sufficient altitude exists, a descent may allow the crew to reestablish visual contact with the formation. If sufficient altitude does not exist, the P* should initiate a climb to provide vertical separation from the flight.

(3) Approach: The lead aircraft must maintain a constant approach angle so other aircraft in the formation will not have to execute excessively steep, shallow, or slow approaches. Aircraft should not descend below the aircraft ahead of them in the formation and entering their rotor-wash. This could result in an N_1 over-speed and/or engine over-temperature, loss of aircraft control, or entering a settling with power condition. In the event an aircraft in the flight loses visual contact with the formation, they will immediately make a radio call to the formation and execute a go-around in the briefed direction.

c. Reorientation procedures:

(1) After announcing the aircraft has a loss of visual contact with the formation, lead will announce and maintain heading, altitude, and airspeed, turning only to avoid known obstacles or enemy threat. Lead will also announce his or her position relative to the next waypoint or rally point. The remainder of the formation will continue to follow lead. The crewmember who has lost visual contact will announce his or her position relative to the same waypoint or rally point to assist in reorientation to the flight. This procedure will continue until the formation is reoriented and joined.

(2) Considerations should include, but are not limited to, rallying to a known point, use of covert/overt lighting, and ground rally. METT-TC, power available, and ambient light will influence how contact is reestablished.

(3) Situations may occur when an aircraft rejoins the flight in another position than briefed. Only after the entire flight is formed can the mission commander proceed with the mission.

d. Aircrew briefing: All multi-aircraft operations will be briefed using a unit approved multi-aircraft/mission briefing CL. Table 4-4 lists mandatory items that must be included in all multi-aircraft briefings.

Table 4-4. Multi-aircraft operations briefing checklist

1.	Formation type(s): takeoff, cruise, and approach.
2.	Altitude.
3.	Airspeed: outbound to release point, cruise, inbound from start point.
4.	Aircraft lighting.
5.	Loss of communications procedures.
6.	Lead change procedures.
7.	Loss of visual contact/in-flight link-up/rally points.
8.	Actions on contact.
9.	IIMC procedures.
10.	Downed aircraft procedures/personnel recovery/CSAR.

NIGHT OR NIGHT VISION GOGGLE CONSIDERATIONS: Closure rates are more difficult to determine. Keep changes in the formation to a minimum. All crewmembers must avoid fixation by using proper scanning techniques. During unaided night/NVG formation flight, the crew should use formation and position lights to aid in maintaining the aircraft's formation position. Lighting will be IAW AR 95-1 and unit SOP.

SNOW/SAND/DUST CONSIDERATIONS:

1. Takeoff: A simultaneous formation takeoff may not be possible due to loss of visual contact with other aircraft in the formation. Crews should consider taking off single ship, then conducting an in-flight link up once clear of the snow/sand/dust cloud. During single-ship takeoff, it is important to notify the formation when clear of the dust cloud to notify the next aircraft ready for takeoff.

2. Approach: A landing should be made to the ground with forward groundspeed and heading for all aircraft off-set by 10 degrees from lead's landing direction. This will ensure lateral separation during periods of degraded visibility. For example, lead lands heading 360 degrees, chalk 2 lands 350 degrees, chalk 3 lands 010 degrees, chalk 4 lands 350 degrees, and chalk 5 lands 010 degrees.

TRAINING AND EVALUATION REQUIREMENTS:

1. Training will be conducted in the aircraft or a Mi-17 FS.
2. Evaluation will be conducted in the aircraft or a Mi-17 FS.

REFERENCES: Appropriate common references and ATTP 3-18.12.

TASK 2012

Perform Tactical Flight Mission Planning

CONDITIONS: Prior to flight in a Mi-17 helicopter or a Mi-17 FS, and given a mission briefing, navigational maps, a navigational computer, Army-approved mission planning station and software (if available), and other flight planning materials as required.

STANDARDS: Appropriate common standards and the following additions/modifications:

1. Analyze the mission using the factors of METT-TC.

2. Operate the Army-approved mission planning station and software, if available.

3. Perform a map/photo reconnaissance using the available map media, photos, and Army-approved mission planning station and software. Ensure all known hazards to terrain flight are plotted.

4. Select the appropriate flight altitudes.

5. Develop load plan and verify aircraft weight and balance (Task 1012).

6. Select appropriate primary and alternate routes and enter all of them on a map, route sketch, or into the Army-approved mission planning station and software if available.

7. Determine the distance ±1 kilometer, ground speed ±5 knots, and ETE ±2 minutes for each leg of the flight.

8. Determine the fuel required and reserve IAW AR 95-1, ±100 liters.

9. Obtain and analyze the weather briefing to determine that weather and environmental conditions are adequate to complete the mission.

10. Load mission data to data transfer cartridge or data load unit, if available.

11. Print out time distance heading cards, waypoint lists, crew cards, communication cards, and kneeboard cards as required.

12. Conduct a thorough crew mission briefing.

DESCRIPTION:

1. Crew actions.
   a. The PC/AMC will delegate mission tasks to crewmembers, have overall responsibility for mission planning, and conduct a thorough crew mission briefing. PC/AMC will analyze the mission in terms of METT-TC.
   b. The P and NCMs will perform the planning tasks directed by the PC/AMC. The P and NCMs will report their planning results to the PC/AMC.

2. Procedures.
   a. Analyze the mission using METT-TC factors.
   b. Conduct a map or aerial photo reconnaissance.
   c. Obtain a thorough weather briefing covering the entire mission; to include sunset and sunrise times, density altitudes, winds, and visibility restrictions. Night mission briefings will include moonset and moonrise times and ambient-light levels, if available.
   d. Determine primary and alternate routes, terrain flight modes, and movement techniques. Determine time, distance, and fuel requirements using the navigational computer or Army-approved mission planning station and software if available.
   e. Annotate the map or Army-approved mission planning station and software, if available, with sufficient information to complete the mission IAW the unit SOP. This includes waypoint coordinates that define the entry routes into the GPS/Army-approved mission planning station and software, if available. Consider such overlay items as hazards, check points, observation posts, and friendly and enemy positions. Review contingency procedures.

Note. Evaluate weather impact on the mission. Considerations should include aircraft performance and limitations.

NIGHT OR NIGHT VISION GOGGLE CONSIDERATIONS: More detailed flight planning is required when the flight is conducted in reduced visibility, at night, or in the NVG environment. FM 3-04.203 contains details about night navigation. NVG navigation with standard maps can be difficult because of map colors, symbology, and colored markers used during map preparation.

TRAINING AND EVALUATION REQUIREMENTS:
1. Training will be conducted academically.
2. Evaluations will be conducted academically.

REFERENCES: Appropriate common references.

TASK 2022

Transmit a Tactical Report

CONDITIONS: In a Mi-17 helicopter, a Mi-17 FS, or academically and given sufficient information to compile a tactical report.

STANDARDS: Appropriate common standards plus transmit the appropriate report using the current SOI.

DESCRIPTION:

1. Crew actions.

    a. The P* and NCMs will focus primarily outside the aircraft to clear the aircraft and provide adequate warning of traffic or obstacles. The P* will announce any maneuver or movement before execution.

    b. The P will assemble and transmit the report. The P will use the correct format, as specified in the SOI, and transmit the report to the appropriate agency. The FE should also be able to transmit the report if the P is unable to do so.

2. Procedures.

    a. Use an established format to report information to save time, minimize confusion, and ensure completeness.

    b. Assemble the report in the correct format and transmit it to the appropriate agency. Standard formats may be found in the SOI or other sources.

Note. Encryption is only required if information is transmitted by non-secure means.

TRAINING AND EVALUATION REQUIREMENTS:

1. Training may be conducted in the aircraft, a Mi-17 FS, or academically.

2. Evaluations will be conducted in the aircraft or academically.

REFERENCES: Appropriate common references, FM 2-0, and the SOI.

TASK 2024

Perform Terrain Flight Navigation

CONDITIONS: In a Mi-17 helicopter or a Mi-17 FS and given a mission briefing and required maps and materials.

STANDARDS: Appropriate common standards and the following additions/modifications:

1. During nap-of-the-earth (NOE) flight, know the en route location within 200 meters.

2. During contour flight or low-level flight, know the en route location within 500 meters.

3. Locate each objective within 100 meters.

4. Arrive at each objective at the planned time, ±2 minutes (if an objective arrival time was given in the mission briefing).

ESCRIPTION:

1. Crew actions.

   a. The P* will remain focused outside the aircraft and respond to navigation instructions and cues given by the P. The P* will acknowledge commands issued by the P for heading and airspeed changes necessary to navigate the desired course. The P* will announce significant terrain features to assist the P in navigation.

   b. The P will furnish the P* with the information required to remain on course. The P will announce all plotted wires/hazards before approaching their location. The P will use rally terms and terrain features to convey instructions to the P*. Examples of these terms are "turn left to your 10o'clock," "stop turn," and "turn down the valley to the left." If using the horizontal situation indicator during low-level flight, 80 feet AHO, the P may include headings. The P should use electronically aided navigation to help arrive at a specific checkpoint, turning point, or objective.

   c. The P*, P, and NCMs should use standardized terms to prevent misinterpretation of information and unnecessary cockpit conversation. The crew must look far enough ahead of the aircraft at all times to assist in avoiding traffic and obstacles.

2. Procedures.

   a. During NOE and contour flight, identify prominent terrain features located some distance ahead of the aircraft and lying along or near the course.

   (1) Using these terrain features to key on, the P* maneuvers the aircraft to take advantage of the terrain and vegetation for concealment.

   (2) If this navigational technique does not apply, identify the desired route by designating a series of successive checkpoints.

   (3) To remain continuously oriented, compare actual terrain features with those on the map.

   (4) An effective technique is to combine the use of terrain features and rally terms when giving directions. This will allow the P* to focus his or her attention outside the aircraft.

   b. For low-level navigation, the time and distance can be computed effectively. This means the P* can fly specific headings and airspeeds. Each of the methods for stating heading information is appropriate under specific conditions.

   (1) When a number of terrain features are visible and prominent enough for the P* to recognize them, the most appropriate method is navigation instruction toward a terrain feature in view.

   (2) When forward visibility is restricted and frequent changes are necessary, controlled turning instructions are more appropriate.

   (3) Clock headings are recommended when associated with a terrain feature and with controlled turning instructions.

Note. For additional information, refer to Tasks 1044 and 1046.

> *Note.* The aircrew should incorporate the use of Army-approved mission planning station and software, if available, with this task.

NIGHT OR NIGHT VISION GOGGLE CONSIDERATIONS:

1. Conducting the flight in reduced visibility (or at night) requires more detailed and extensive flight planning and map preparation. FM 3-04.203 contains details on night navigation. NVG navigation with standard maps can be difficult because of map colors, symbology, and colored markers used during map preparation.

2. Use proper scanning techniques to ensure obstacle avoidance.

TRAINING AND EVALUATION REQUIREMENTS:

1. Training may be conducted in the aircraft or a Mi-17 FS.

2. Evaluations will be conducted in the aircraft.

REFERENCES: Appropriate common references and FM 3-25.26.

TASK 2026

Perform Terrain Flight

CONDITIONS: In a Mi-17 helicopter or a Mi-17 FS, with tactical flight mission planning completed.

STANDARDS: Appropriate common standards and the following additions/modifications:
1. Maintain altitude and airspeed appropriate for the selected mode of flight, visibility, and METT-TC.
2. Maintain aircraft in trim during contour and low-level flight.

DESCRIPTION:
1. Crew actions.
   a. The P* will focus primarily outside the aircraft and acknowledge all navigational and obstacle-clearance instructions given by the P/NCMs. The P* will announce the intended direction of flight or any deviation from instructions given by the P. During terrain flight, the P* is primarily concerned with threat and obstacle avoidance.
   b. During terrain flight, the P* is primarily concerned with threat and obstacle avoidance.
   c. The P will provide adequate warning to avoid obstacles detected in the flight path or identified on the map. The P and NCMs will assist in clearing the aircraft and provide adequate warning of obstacles, unusual attitudes, altitude changes, or threat. The P and NCMs will announce when their attention is focused inside the aircraft and when attention is reestablished outside.
   d. During contour flight, the P/NCMs will advise the P* whenever an unannounced descent is detected. If the descent continues without acknowledgement or corrective action, the P will again advise the P* and be prepared to make a collective-lever control input. The P will raise the collective-lever when it is apparent the aircraft will descend below 25 feet AHO.
   e. During NOE flight, the P/NCMs will advise the P* whenever an unannounced descent is detected. The P will immediately raise the collective-lever when it is apparent the P* is not taking corrective action and the aircraft will descend below 10 feet AHO.

2. Procedures. Terrain flight is close to the earth's surface. The modes of terrain flight are NOE, contour, and low-level. Crewmembers will seldom perform pure NOE or contour flight. Instead, they will alternate techniques while maneuvering over the desired route.
   a. NOE flight. Perform NOE flight at varying airspeeds and altitudes as close to the earth's surface as vegetation, obstacles, and ambient light permit.
   b. Contour flight. Perform contour flight by varying altitude and while maintaining a relatively constant airspeed, depending on the vegetation, obstacles, and ambient light. Generally, follow the contours of the earth.
   c. Low-level flight. Perform low-level flight at a constant airspeed and altitude. To prevent or reduce the chance of detection by enemy forces, fly at the minimum safe altitude that will allow a constant altitude.

Note. Performing this maneuver in certain environments may require hover OGE power. Evaluate each situation for power required versus power available.

Note. Terrain flight is considered sustained flight below 200 feet AGL, except during takeoff and landing.

Note. The aircrew should incorporate the use of Army-approved mission planning station and software, if available, with this task.

NIGHT OR NIGHT VISION GOGGLE CONSIDERATIONS:
1. Wires are difficult to detect with NVG.
2. Use proper scanning techniques to ensure obstacle avoidance.

OVERWATER CONSIDERATIONS:

1.   All crewmembers will wear floatation devices IAW AR 95-1.

2.   Overwater flight, at any altitude, is characterized by a lack of visual cues; therefore, it has the potential of causing visual illusions. To minimize spatial disorientation, the crew should use radar altitude hold during overwater flight.

3.   Be alert to any unannounced changes in the flight profile and be prepared to take immediate corrective actions. The radar altimeter low bug should be set to assist in altitude control.

4.   Operations become increasingly more hazardous as references are reduced (open water versus a small lake), water state increases (calm to chop to breaking condition with increasing wave height), and visibility decreases (horizon becomes same color as water, water spray [or rain] on windshield, sunny midday versus twilight).

5.   Hazards to flight such as harbor lights, buoys, wires, and birds must be considered during overwater flight.

TRAINING AND EVALUATION REQUIREMENTS:

1.   Training may be conducted in the aircraft or a Mi-17 FS.

2.   Evaluations will be conducted in the aircraft.

REFERENCES: Appropriate common references and FM 3-25.26.

TASK 2036

Perform Terrain Flight Deceleration

CONDITIONS: In a Mi-17 helicopter or a Mi-17 FS.

STANDARDS: Appropriate common standards and the following additions/modifications:
1. Maintain heading alignment with the selected flight path, ±10 degrees.
2. Maintain the tail rotor clear of all obstacles.
3. Decelerate to the desired airspeed or to a full stop at the selected location, ±50 feet.

DESCRIPTION:
1. Crew actions.
   a. The P* will focus primarily outside the aircraft to clear the aircraft throughout the maneuver. The P* will announce their intention to decelerate or come to a full stop, any deviation from the maneuver, and completion of the maneuver.
   b. The P and NCMs will provide adequate warning to avoid obstacles detected in the flight path, announce when their attention is focused inside the cockpit, and announce when attention is reestablished outside.
2. Procedures.
   a. The P* will initially raise the collective to maintain the altitude of the tail rotor. (Collective control application may not be necessary when initiation of the maneuver is at higher airspeeds.)
   b. The P* must consider variations in the terrain and obstacles when determining tail rotor clearance. The P* will apply aft cyclic to slow to the desired airspeed (or come to a full stop) while adjusting the collective to maintain the altitude of the tail rotor.
   c. The P* will maintain heading with the pedals and will make all control movements smoothly. If the altitude of the tail rotor increases during the deceleration, the P* may need to lower the collective to return to the desired altitude.
   d. If the aircraft attitude is changed excessively or abruptly, it may be difficult to return the aircraft to a level attitude and over controlling may result.

Note. Performing this maneuver in certain environments may require hover OGE power. Evaluate each situation for power required versus power available.

NIGHT OR NIGHT VISION GOGGLE CONSIDERATIONS: The P* must avoid making abrupt changes in aircraft attitude as the NVG will limit the field of view. The P* should maintain proper scanning techniques to ensure obstacle avoidance and clearance.

TRAINING AND EVALUATION REQUIRMENTS:
1. Training will be conducted in the aircraft or a Mi-17 FS.
2. Evaluation will be conducted in the aircraft.

REFERENCES: Appropriate common references and FM 3-25.26.

TASK 2042

Perform Actions on Contact

CONDITIONS: In a Mi-17 helicopter, a Mi-17 FS, or academically.

STANDARDS: Appropriate common standards and using the correct actions on contact consistent with the tactical situation.

1. If appropriate, immediately deploy to a covered and concealed position using suppressive fire.
2. Continue observation as appropriate for the mission.
3. Transmit tactical report IAW SOI, the unit SOP or mission briefing.

DESCRIPTION:

1. Crew actions. When engaged by or upon detecting the enemy, the crewmember identifying the threat will announce the nature (visual observation, radar detection, or hostile fire) and direction of the threat.

   a. Proper pre-mission planning and intelligence data may aid in developing flight profiles and route selection to avoid hostile fire.

   b. Fly the helicopter to a concealed area using the evasive techniques below and suppressive fire, as required. Choose a course of action supporting the mission and the intent of the unit commander's directives.

   c. If engaged by the enemy, the crew will announce the nature (hostile fire or radar detection) and direction of the threat. The crewmember first identifying the threat is responsible for announcing the threat bearing, relative to the aircraft, and launching countermeasures/suppressive fire as necessary.

   d. The P* will announce the direction of flight to deploy to cover and remain focused outside the aircraft during the evasive maneuver and clearing.

   e. Avoid over-controlling/excessive maneuvering that may result in loss of aircraft control (or insufficient power) to recover from the maneuver.

   f. The P and NCMs will remain focused primarily outside the aircraft and announce adequate warning to avoid obstacles detected during the evasive maneuver.

   g. The P will remain oriented on the threat's location. The P will announce warnings to avoid obstacles when his or her attention is focused inside the aircraft, again when his or her attention is reestablished outside, and will transmit a tactical report.

   h. The NCMs will remain focused primarily outside the aircraft and announce adequate warning to avoid obstacles. They will also provide suppressive fire as required.

   Note. The FE must be able to transmit a tactical report IAW the SOI, unit SOP, or mission briefing.

   ---

2. Procedures. The specific maneuver required will depend on the type of hostile fire encountered. The guidance below may assist with developing actions on contact for the given threat system. A thorough intelligence briefing will help to identify the actions on contact the crew can expect to take for the most probable threat system employment.

   a. Tanks, rocket propelled grenade, and small arms.

   (1) If concealment is available, deploy toward the area of concealment.

   (2) If concealment is not readily available, immediately turn to an oblique angle while applying forward cyclic. Turn to an oblique angle from the hostile fire to minimize the aircraft's profile and make it a more difficult target. Apply forward cyclic to accelerate while descending in an attempt to mask the aircraft. Make turns of unequal magnitude, at unequal intervals, and small altitude changes to provide the best protection until beyond the effective range of hostile weapons.

   (3) If the situation permits, employ immediate suppressive fire.

   b. Large caliber, anti-aircraft fire (radar-controlled).

   (1) Execute an immediate 90-degree turn and mask the helicopter.

(2) After turning, **do not** maintain a straight line of flight or the same altitude for more than 10 seconds before initiating a second 90-degree turn.

(3) To reduce the danger, descend immediately to NOE altitude.

c.   Fighters.

(1) On sighting a fighter, try to mask the helicopter.

(2) If the fighter is alone and executes a dive, turn the helicopter toward the attacker and descend. This maneuver will cause the fighter pilot to increase the attack angle.

(3) Depending on the fighter's dive angle, it may be advantageous to turn sharply and maneuver away once the attacker is committed. The fighter pilot will then have to break off the attack to recover from the maneuver.

(4) Once the fighter breaks off the attack, maneuver the helicopter to take advantage of terrain, vegetation, and shadow for concealment.

d.   Heat-seeking missiles.

(1) As appropriate, employ the ASE to counter heat-seeking devices while maneuvering to avoid the threat. If a missile is detected, apply forward cyclic and turn the heat source away from the threat. Attempt to mask the aircraft while orienting crew-served weapons for suppressive fire.

(2) MAWS-If a missile is detected, initially maintain course/altitude and allow the countermeasure system to defeat the threat. Perform the appropriate combat maneuvering flight (Task 2127) maneuver and turn to an oblique angle from the threat to minimize the profile of the aircraft while evading. Delay a descent momentarily after last flare launch to allow for IR missile decoy. Attempt to mask the aircraft while orienting/employing crew served weapons for suppressive fire.

e.   Radar-guided missiles. Perform the appropriate combat maneuvering flight (Task 2127) maneuver to break the line of sight to the radar source while simultaneously activating chaff if available. Maneuver away from the threat source and attempt to keep the threat system to the right rear or left rear of aircraft and simultaneously dispense chaff. Attempt to keep the chaff cloud between the aircraft and the threat source. Once chaff is dispensed, turn the aircraft to maneuver away from the chaff cloud and continue to chaff and turn until the aircraft is masked.

f.   Antitank-guided missiles. Some missiles fly relatively slowly and are avoidable by rapidly repositioning the helicopter. If terrain or vegetation is unavailable for masking, remain oriented on the missile as it approaches. As the missile is about to impact, rapidly change flight path or altitude to evade it.

g.   Artillery. Depart the impact area, and determine CBRNE requirements.

Note. Dispensing chaff while maneuvering may cause tracking radars to break lock.

---

h.   After successfully deploying to cover, the crew will—

(1) Report the situation.

(2) Develop the situation.

(3) Choose a course of action, if not directed by the unit commander. (The P*/P will announce the unit commander's directive if not monitored by the other crewmember.)

i.   If hit by hostile fire, rapidly assess the situation and determine an appropriate course of action.

(1) Assess aircraft controllability.

(2) Check all instruments and warning/caution lights. If a malfunction is indicated, initiate the appropriate emergency procedure.

(3) If continued flight is possible, take evasive action.

(4) Radio call your situation, location, action, and request for assistance if desired.

(5) Continue to be alert for unusual control responses, noises, and vibrations.

(6) Monitor all instruments for an indication of a malfunction.

(7) After landing, inspect the aircraft to determine the extent of damage and if flight can be continued.

Note. Proper employment of terrain flight techniques will reduce exposure to enemy threat weapon systems.

Note. Threat elements will be harder to detect. Rapid evasive maneuvers will be more hazardous due to division of attention and limited visibility. Maintain SA with regard to threat and hazard location.

Note. Performing this maneuver in certain environments may require hover OGE power. Evaluate each situation for power required versus power available.

NIGHT OR NIGHT VISION GOGGLE CONSIDERATIONS:

1. At low ambient light levels, obstacle detection is difficult. The P* may experience spatial disorientation if he or she executes abrupt maneuvers. Proper scanning techniques and good cockpit communication are necessary to avoid these hazards.

2. The crew should consider using artificial lighting if the ambient light level is insufficient for obstacle detection.

TRAINING AND EVALUATION REQUIREMENTS:

1. Training may be conducted in the aircraft, a Mi-17 FS, or academically.

2. Evaluations will be conducted in the aircraft, a Mi-17 FS, or academically.

REFERENCES: Appropriate common references, and ASET.

TASK 2048

Perform External (Sling) Load Operations

CAUTION

A static electricity discharge wand will be utilized IAW TM 4-48.09.

CONDITIONS: In a Mi-17 helicopter or a Mi-17 FS, with the crew briefing completed, aircraft cleared, and a certified load.

STANDARDS: Appropriate common standards and the following additions/modifications:

1. RCM.
   a. Ensure the aircraft remains clear of the load and any obstacles.
   b. Perform a vertical ascent with the load to a load height of 10 feet, ±3 feet, AGL.
   c. Perform a vertical descent with the load to the desired touchdown point, ±5 feet.
   d. Ensure the load remains clear of any obstacles and is not dropped or dragged.

2. NCM.
   a. Direct the P* over the load for hookup using no more than two directions at a time.
   b. Properly direct the P* to the release area and clear the load for release.

DESCRIPTION:

1. Crew actions.
   a. The PC will conduct a thorough crew briefing and ensure that all crewmembers are familiar with external load operations, emergency, and communication procedures. The PC will ensure that the external load is certified (for example, DA Form 7382 [Sling Load Inspection Record] or foreign equivalent). The PC will determine the direction for takeoff by analyzing the tactical situation, the wind, the long axis of the take-off area, and the lowest obstacles and will confirm that required power is available by comparing the information from the PPC to the hover power check.
   b. The P* will remain focused outside the aircraft throughout the maneuver. He or she will monitor altitude and avoid obstacles.
   c. The P and/or FE will monitor the cockpit instruments and assist the P* in clearing the aircraft. The P will make all radio calls and perform the hover power check. The FE performs the before takeoff check.
   d. The P and/or CE not calling the load will assist in clearing the aircraft and provide adequate warning of obstacles.
   e. The CE calling the load will attach his or her restraining harness to a tie-down ring.
   f. The CE calling the load will remain primarily focused on the load. The CE will guide the P* during the load pickup, advise of the load condition in flight, and direct the P* when setting down the load.

2. Procedures.
   a. Hookup and hover.

      (1) The P/FE may place the radio switches on the P* ICS "off" as directed by the P*.

      (2) The P* will announce when the load is under the rotor system or when he or she loses sight of the load. The P* will follow hand signals from the signalman or commands from the NCM calling the load to hover over the load. The P* will remain vertically clear of and centered over the load.

      (3) The cabin NCM calling the load will direct the P* over the load using no more than two directions at a time. He or she will advise the P* when the load is hooked, remove slack from the sling, and ascend vertically to a stabilized load height of 10 feet.

      (4) If a ground crew is used for hookup, the cabin NCM calling the load will advise the P* when and in what direction the crew cleared the load and the aircraft. He or she will monitor the load rigging and advise the P* when the slings are tight. During the load hookup and after the slings are tight, the P

and/or FE should refer to the radar altimeter for actual aircraft height AGL. The P and/or FE should then round up the height to the nearest 5 feet and add 10 feet for the appropriate hover height.

(5) The cabin NCM calling the load will call out load height in 1-foot increments until the load is 10 feet off the ground. When the load is stable and the rigging appears safe, the cabin NCM will announce the load is cleared for flight.

b. Takeoff.

(1) The P* will maintain a 10-foot load height until the P completes a hover power check and the FE completes a before-takeoff check.

(2) Before takeoff, the P* will ensure that the load is cleared for flight by the cabin NCM calling the load. The P* will make smooth control inputs to initiate takeoff and establish a constant angle of climb that will permit safe obstacle clearance.

(3) During takeoff, the cabin NCM will call the aircraft load height AHO at 15 feet, 20 feet, 25 feet, 50 feet, 75 feet, and 100 feet. The P and/or FE will back-up the cabin NCM by calling out the load height by referencing the radar altimeter and announce the altitude if the cabin NCMs altitudes are in error.

(4) When above 100 feet AHO or when clear of obstacles, the P* will adjust attitude and power, as required, to establish the desired rate of climb and airspeed. During the acceleration, he or she will avoid unnecessary nose-low attitudes and over controlling to reduce load oscillation.

(5) The cabin NCM will announce load condition (such as load clear of all barriers, load is stable, and so forth). When aircraft load height is above 100 feet AGL or when clear of obstacles, the P* will increase airspeed slowly to determine the flight characteristics of the load and smoothly adjust flight controls to avoid oscillation.

Note. Performing this maneuver in certain environments may require hover OGE power. Evaluate each situation for power required versus power available.

Note. If load oscillation develops, the primary method for arresting the oscillation is to decrease airspeed. Additional measures may include shallow turns or banks, small climbs or descents, or a combination of any or all methods.

c. En route.

(1) The P/FE will turn on the P*'s radio switches as required.

(2) The P will advise the P* to make smooth control applications to prevent load oscillation. The cabin NCM will monitor the load for oscillation/load height and advise the P* of the status of the load.

d. Approach and load release.

(1) The P/FE may turn off the P*'s radio switches as directed.

(2) The P* will establish and maintain an approach angle that will keep the load clear of obstacles to the desired point of termination.

(3) The P* will establish a rate of closure appropriate for the conditions of the load. (A go-around should be made before descending below obstacles or decelerating below ETL.)

(4) The cabin NCM calling the load will call the aircraft load height altitude AHO on approach at 100 feet, 75 feet, 50 feet, 25 feet, 20 feet, 15 feet, and 10 feet. The P and/or FE will back up the cabin NCM calling the load by referencing the radar altimeter and announce the altitude if the cabin NCMs altitudes are in error.

(5) The P* will terminate the approach at a stationary hover with the load 10 feet above the intended release point. The cabin NCM will confirm that the release point is clear and direct the P* to the release point with no more than two directions at one time. The cabin NCM will then clear the load down vertically; he or she will call out load height in 1-foot increments until the load is completely on the ground.

(6) The cabin NCM will continue to call descent to obtain slack in the slings, and then hover laterally to ensure the clevis is clear of the load before releasing the load. The cabin NCM will advise the P* when the clevis is clear.

(7) The cabin NCM will release the load upon confirmation from the P* or per the unit SOP. The cabin NCM will confirm the load is released before clearing the P* to reposition from the release point.

Note. Before conducting an external load operation, all crewmembers must ensure that they are able to communicate with each other.

Note. The P* will not allow the external load to descend below the hover height until the cabin NCM calling the load has cleared the load to the ground.

Note. Loads will meet external air transportability (EAT) requirements IAW FM 4-20.197 or foreign equivalent. Procedures for air transportation of hazardous materials will be IAW AR 95-27 or foreign equivalent.

Note. If possible, avoid flight over populated areas.

Note. Before the mission, the PC will ensure all crewmembers and the hookup crew are familiar with hand and arm signals contained in FM 21-60 and with forced landing procedures. In the event of a forced landing, the aviator should land the aircraft to the right of the load.

Note. In the event of mission requirements for a hook-up from a hover, the winch may be configured through the cargo hatch IAW the flight manual. A second NCM will be positioned at the cargo hatch for hook up only.

NIGHT OR NIGHT VISION GOGGLE CONSIDERATIONS:

1. For unaided night flight, both searchlights should be operational. If a NVG filter is installed, it should be removed.

2. When NVG are used, hovering with minimum drift is difficult and requires proper scanning techniques and crew coordination. If possible, use an area with adequate ground contrast and reference points. Visual obstacles, such as shadows, should be treated the same as physical obstacles.

3. The cabin NCM should wear NVG during NVG external load operations. A flashlight with a NVG compatible lens may be used to view the load. The NCM will notify the PC anytime he or she must flip up the NVG; white lighting may be used as necessary to view the hook or load.

4. During load hookup and after the slings are tight, the P/FE should refer to the radar altimeter for actual aircraft height AGL. The P and/or FE should round up height to the nearest 5 feet and add 10 feet for the appropriate hover height

5. During the approach, the P and/or FE should monitor the radar altimeter from 100 feet to the hover height obtained in paragraph 4. The P and/or FE will call out the altitude in increments of 25 feet down to the sling altitude. The cabin NCM will monitor the load and inform the P* if it is determined that the rate of descent or airspeed is excessive or if the accuracy of the radar altimeter is in doubt. When the P/FE announces the sling altitude, the cabin NCM will clear the load down to the release area, calling out altitude in 1-foot increments to the ground.

6. A second NCM is required in the cabin area for airspace surveillance.

TRAINING AND EVALUATION REQUIREMENTS:

1. Training will be conducted in the aircraft or a Mi-17 FS.

2. Evaluation will be conducted in the aircraft.

REFERENCES: Appropriate common references, AR 95-27, FM 4-20.197, FM 4-20.198, FM 4-20.199, FM 21-60, and TM 10-1670-295-23&P.

TASK 2052

Perform Water Bucket Operations

> **WARNING**
>
> Never dump water onto ground personnel, as the water impact could result in injury. Minimize hovering or flying slowly over fires. Rotor wash fans the flames, which may cause more hazards to ground crews. When performing this task with cabin doors open, ensure any personnel in the cabin area are wearing a safety harness secured to a tie-down ring or are sitting in a seat with seat belt fastened.

Note. The water bucket, when loaded, is a high-density load with favorable flight characteristics. Reduced V_{NE} and bank angle limits must be kept in mind. Much of the mission profile is flown at high GWT and low airspeed. In addition, density altitude is greatly increased in the vicinity of a major fire. Performance planning must receive special emphasis.

CONDITIONS: In a MI-17 helicopter, with an operational cargo hook, water bucket, required briefings, checks completed, and an AWR.

STANDARDS: Appropriate common standards and the following additions/modifications:

1. RCM.
    a. Conduct permission planning to determine fuel and bucket cinching requirements. Verify the aircraft will remain within GWT and CG limitations for the duration of the flight.
    b. Conduct a thorough crew briefing.
    c. In conjunction with the NCMs, complete the required checks to ensure proper system operation before mission departure.
    d. Operate the water bucket system according to manufacturer specifications.
    e. Recognize and respond to a water bucket system malfunction.
    f. Use proper dipping procedures for the water bucket type.
    g. Demonstrate knowledge of fire behavior and terminology.
    h. Hookup and hover.
        (1) Maintain vertical ascent heading, ±10 degrees.
        (2) Maintain altitude of load, ± 10feet AGL, +3 foot.
        (3) Complete hover power checks.
    i. En route, maintain load obstacle clearance (minimum 50 feet AHO).
    j. Approach and water release.
        (1) Evaluate fire/simulated fire for flight path and altitude requirements.
        (2) Maintain a constant approach angle to ensure the load safely clears obstacles.
        (3) Maintain ground track alignment with selected approach path.
        (4) Execute a smooth and controlled pass or termination over the intended point/area of water drop.
        (5) Deploy water as directed in proper location, orientation, and/or length.

2. NCM. In conjunction with RCMs, complete required water bucket checks to ensure proper system operation before mission departure and attach water bucket to the aircraft.
    a. Ensure the water bucket is configured for the condition and mode of flight.
    b. Recognize and respond to a water bucket system malfunction.
    c. Demonstrate knowledge of fire behavior and terminology.

DESCRIPTION:

1. Crew actions.

   a. The PC will conduct a thorough crew, external load, and water bucket briefing. The PC will ensure all crewmembers are familiar with water bucket operations and emergency/communication procedures. He or she will ensure a DA Form 7382 has been completed. The PC will confirm required power is available by comparing the information from the PPC to the hover power check.

   b. The P* will remain focused primarily outside the aircraft throughout the maneuver. He or she will monitor altitude and avoid obstacles.

   c. The P will monitor the cockpit instruments and assist the P* in clearing the aircraft. The P will set cargo hook switches, as required, and should make all radio calls. When directed by the P* during the approach, the P will place the cargo hook master switch to the "armed" position. The NCM will release the water IAW the crew briefing.

   d. The P and NCM will assist in clearing the aircraft and provide adequate warning of obstacles. They will announce when their attention is focused inside the aircraft and when attention is reestablished outside. The NCM will remain focused primarily on the bucket. The NCM will guide the P* during the bucket pickup, advise of the bucket condition in flight, provide directions and assistance when the water is dumped, and direct the P* when setting down the bucket.

   e. The NCM will advise the P* of any water bucket faults or failures.

   f. External load procedures IAW Task 1063 will be used for normal external load techniques and load call-outs. The NCM will advise the P* when the water bucket is in the water, filling, full, water deploying, and empty. The NCM will instruct the P* as necessary to keep the electrical attachment assembly from entering the water.

2. Procedures. Crewmembers will follow the water bucket guide provided in table 4-5.

Table 4-5. Sample water bucket guide

Water bucket preflight check
1. Bottom of chain for tears in fabric.
2. Shackle and lockwire or tie-wrap condition.
3. Sidewall battens.
4. Diagonal M-straps connecting the suspension cables for wear.
5. Purse lines on the fabric dump valve.
6. Cinch strap belt-the end opposite the D-ring shall not have a knot.
7. Suspension lines for frays, kinks, and knots.
8. Ballast pouch in the bucket for rips or holes.
9. Control head for secure fittings.
10. Tripline for kinks, frays, or loose swages.
11. Perform operational check of the control head.
Dumping Water
1. Pilot calls -"altitude, airspeed, and monitors radar altimeter during pass.
2. NCM calls "prepare to open bucket/doors" approximately 10 seconds from target.
3. NCM calls over target "open bucket/doors."
4. NMCs respond "bucket/doors open, bucket is ¾, ½, ¼, bucket empty."
Note. Water bucket doors are open or closed depending on bucket type and clear for flight, as required.
Landing
1. Normal load approach.
2. Clear bucket to ground.
3. Clear to slide (direction) away from load.
4. Release the slings and disconnect electrical lines.
5. Recover bucket and secure in aircraft.

Table 4-5. Sample water bucket guide (cont.)

1. Open the bucket, if necessary.
2. Call bucket open, bucket empty.
3. Jettison the load, if necessary.
4. Call load jettisoned.
5. Hook operations-normal and emergency.
6. Lost communication procedures.

    a.   Lost communication procedures.

    b.   Preflight.

       (1) The PC will analyze the mission using METT-TC and determine the amount of water required to conduct the mission and the initial profile to be used during the water emplacement.

       (2) The NCMs will ensure the water bucket is installed and all installation checks are completed according to the unit SOP.

       (3) The crew will conduct ground checks IAW the manufacturer's procedures to confirm the proper operation of the water bucket before takeoff.

    c.   Hook-up and hover.

       (1) Once the water bucket is placed on the ground beside the aircraft and all associated wiring is installed, place the cargo hook master switch in the "arm" position.

       (2) Follow verbal signals from the NCM to hover over the water bucket. Apply control movements as necessary to remain vertically clear and centered over the water bucket.

       (3) Once in this position, smoothly apply collective input until all slack is removed from the suspension cable. Maintain heading with pedals.

       (4) Apply additional collective to raise the bucket to 10 feet AGL. Monitor aircraft instruments to ensure aircraft limitations are not exceeded.

    d.   Water pickup. Evaluation of the water pickup should include depth, obstacles, water current, and availability of hover references.

       (1) Bambi bucket water pickup.

          (a) Arrive over water source with no forward ground speed and a bucket height of 10 feet above water level.

          (b) Slowly reduce the collective and apply a slight amount of forward cyclic until the Bambi bucket contacts the water. Follow the NCM's verbal guidance to remain centered over the bucket as it fills, applying cyclic, collective, and pedals as necessary.

          (c) The pilot can vary the bucket's capacity by varying the speed at which it is pulled from the water. A slow lift gives minimum fill. A fast lift gives maximum fill.

          (d) When the NCM indicates the bucket is ready (or full), increase the collective lever until all slack is removed from the suspension cable and the lip of the bucket is clear of the water; maintain heading with pedals.

          (e) Apply additional collective to raise the filled bucket clear of the water's surface to a height of 10 feet. Ensure the bucket is holding the water and monitor aircraft instruments to ensure aircraft limitations are not exceeded.

       (2) Sims and simplex water pickup.

          (a) Arrive over water source with no forward ground speed and a bucket height of 10 feet above water level.

          (b) Ensure the bucket doors are open.

          (c) Slowly reduce the collective until the bucket makes contact with the water. Once the bucket has submerged in the water, follow the NCM's verbal guidance to remain centered over the bucket as it fills, applying cyclic, collective, and pedals as necessary.

(d) When the NCM indicates the bucket is full, he or she will close the bucket doors and ensure the bucket is ready.

(e) The P* can increase collective until all slack is removed from the suspension cable and the lip of the bucket is clear of the water. Maintain heading with pedals.

(f) Apply additional collective to raise the filled bucket clear to the water's surface to a height of 10 feet. Ensure the bucket is holding the water and monitor aircraft instruments to ensure aircraft limitations are not exceeded.

Note. Use the manufacturer's recommended en route airspeeds for each type of water bucket. This prevents the buckets from twisting and pinching the cables.

e.   Takeoff. Establish a constant angle of climb that will permit safe obstacle clearance. When above 100 feet AGL or when clear of obstacles, adjust attitude and power as required to establish the desired rate of climb and airspeed. Smoothly adjust flight controls to prevent bucket oscillation.

Note. Ensure the cargo hook master switch is in the "armed" position when operating at altitudes below 300 feet AHO and in the "off" position above 300 feet AHO.

f.   En route. Maintain the desired altitude, flight path, and airspeed. Make smooth control applications to prevent bucket oscillation. If an oscillation occurs, perform the same procedures as in FM 3-04.203, paragraph 2-70.

g.   Approach and water release.

(1) The PC will determine the most appropriate height and speed for the pattern desired, or IAW the mission briefing.

(2) Evaluation of the fire should include wind direction, velocity, terrain, and type of fire. Fires usually require a drop height of 100 to 200 feet AGL and a ground speed of 30 to 60 knots.

(3) The aircraft's ground track should be upwind and adjusted so the spray will provide maximum cooling to hot spots, as well as dampen unburned vegetation. Altitude and airspeed may be adjusted for fires of varying intensity and types. However, it must be noted that low, slow passes may tend to increase the fire's intensity due to rotor downwash.

(4) When the approach angle is intercepted, decrease the collective lever to establish the descent. When passing below 300 feet AGL, place cargo hook master switch in the ARM position. When reaching the desired airspeed and altitude, the recommended crew coordination terms for bucket operations are as follows:

(a) Approaching the target-"prepare to open the doors" (approximately 10 seconds out).

(b) Over the target-"open doors."

(c) When the drop is complete-"close doors."

h.   Postmission. Ensure the water bucket is serviceable, de-rig aircraft and water bucket, and ensure all documentation is complete on water bucket usage and inspection.

Note. The NCM will advise the P* of the condition of the bucket and call out the water level while releasing water. The bucket manufacturer does not recommend dumping at airspeeds above 50 KIAS.

Note. There is a delay of approximately 0.5 to 1.0 second between the activation of the dump switch and discharge of the water.

Note. Avoid flight over populated areas.

Note. A go-around should also be initiated if visual contact with the water release area is lost or if a crewmember announces "climb, climb, climb." This phrase will only be used when there is not enough time to give detailed instructions to avoid obstacle.

SAND/DUST/SMOKE CONSIDERATIONS: If during the approach, visual reference with the water release area (or obstacles) is lost, immediately initiate a go-around or ITO as required. Be prepared to transition to instruments. Once VMC are regained, continue with the go-around. (If required, releasing the water reduces the GWT of the aircraft and minimizes power demand.)

MOUNTAINOUS AREA CONSIDERATIONS: During an approach, if sufficient power is unavailable or turbulent conditions or wind shift create an unsafe condition, immediately perform a go-around. (If required, releasing the water reduces the GWT of the aircraft and minimizes power demand.)

OVERWATER CONSIDERATIONS:

1. All crewmembers will wear floatation devices IAW AR 95-1.

2. Overwater flight, at any altitude, is characterized by a lack of visual cues and; therefore, it has the potential of causing visual illusions. To minimize spatial disorientation, the crew should use radar altitude hold during overwater flight.

3. Be alert to any unannounced changes in the flight profile and prepared to take immediate corrective actions. The radar altimeter low bug should be set to assist in altitude control.

4. Operations become increasingly more hazardous as references are reduced (open water versus a small lake), water state increases (calm to chop to breaking condition with increasing wave height), and visibility decreases (horizon becomes same color as water, water spray [or rain] on windshield, sunny midday versus twilight).

5. Hazards to flight such as harbor lights, buoys, wires, and birds must be considered during overwater flight.

NIGHT OR NIGHT VISION GOGGLE CONSIDERATIONS (NOT RECOMMENDED):

1. During water bucket operations, the P*'s attention will be divided between the aircraft instruments (altitude and ground speed) and the outside. It is critical during NVG operations that the crewmembers' focus be primarily outside to provide warning to the P* of obstacles (or hazards) during the entire operation.

2. Spatial disorientation can be overwhelming during overwater operations at night. Proper scanning techniques are necessary to avoid spatial disorientation.

TRAINING AND EVALUATION REQUIREMENTS:

1. Training will be conducted in the aircraft.

2. Evaluation will be conducted in the aircraft.

REFERENCES: Appropriate common references, AR 70-62, FM 4-20.197, and water bucket AWR.

TASK 2060

Perform Rescue-Hoist/Winch Operations

WARNING

Ensure that crewmembers in the cabin area are wearing a safety harness secured to a tie-down ring anytime the cabin door is open. The crewmember riding the hoist will be secured either to the aircraft or to the jungle penetrator. Ensure that the cable touches the ground or the water before ground personnel touch the cable. The cable will be charged with in excess of 300,000 volts of static electricity.

CAUTION

Care must be taken not to snag terrain features or foliage with the rescue-hoist cable. This may result in exceeding the structural limitation of the overhead pulley support.

CONDITIONS: In a Mi-17 helicopter equipped with a rescue-hoist/winch system.

STANDARDS: Appropriate common standards and the following additions/modifications:
1. RCM.
   a. Conduct a thorough crew and passenger safety briefing.
   b. Perform rescue-hoist procedures IAW the flight manual.
   c. Perform rescue-hoist/winch procedures IAW the flight manual, unit SOP, FM 8-10-6, and FM 3-04.203.
   d. Maintain appropriate hover altitude, ±5 feet.
   e. Do not allow drift to exceed ±5 feet from the intended hover point.
2. NCM.
   a. Perform a preflight inspection of the rescue-hoist/winch IAW the flight manual and unit SOP.
   b. Ensure the crew, passengers, cargo, and mission equipment are secured.
   c. Operate the rescue-hoist/winch.

DESCRIPTION:
1. Crew actions. Rescue hoist operations.
   a. The PC will conduct a thorough crew briefing and ensure all crewmembers are familiar with rescue-hoist operations, emergency procedures, communication procedures, lowering the flight medic, and lifting the patient off the ground using the hoist or aircraft. The PC will also ensure all crewmembers understand "cut cable" procedures.
   b. The P* will remain focused primarily outside the aircraft throughout the maneuver to ensure aircraft control and obstacle avoidance. The P* will announce the intended point of hover and remain centered over the target, incorporating corrections from the NCM.
   c. The P/FE and NCM will assist in clearing the aircraft and provide adequate warning of obstacles. They will also assist the P* in maintaining a stable hover by providing the P* with information regarding the drift of the aircraft. The P will monitor cockpit indications.
   d. The NCM will ensure the hoist is configured and all lifting devices (such as Jungle penetrator, SKED/Stokes litter, and survivor's slings) are secured in the aircraft before takeoff.
   e. The NCM will ensure the winch is configured for rescue-hoist operations and the appropriate write-up is entered on DA Form 2408-13-1 for the mid-hook being removed.

f. The NCM will conduct the hoist operation IAW FM 3-04.203, the flight manual, and the unit SOP.

2. Procedures.

a. General recovery procedures over land.

(1) Crewmembers alerted approximately 5 minutes before arrival at pickup site.

(2) Crewmembers complete all required checks (such as rescue-hoist control panel switches set, hoist circuit breakers set, ICS selector switches set, and crewmembers reposition for hoist operations).

(3) Make the approach into the wind if possible and plan to terminate the approach at an altitude that will clear the highest obstacle.

(4) Select an appropriate reference point to maintain heading and position over the ground. Once stabilized over pickup site, perform hoist operations IAW FM 8-10-6, FM 3-04.203, the flight manual, and the unit SOP.

b. Inert patient recovery.

(1) General format is the same as over land except the NCM/medical officer (MO) is lowered on the hoist and secures the patient to the recovery device.

(2) Before deploying, all crewmembers will be briefed on the method of recovery (simultaneous or singular recovery of the patient and MO) and a radio communications check should be made between the pilot and NCM/MO.

c. General recovery procedures overwater.

(1) General format is the same as over land except a smoke device may be used to determine wind direction and velocity. Terminate the approach at a 100-foot hover, 20 feet before reaching the patient. Deploy the recovery device and allow it to contact the water before reaching the patient.

(2) All crewmembers will wear floatation devices. Operations become increasingly more hazardous as references are reduced (open water versus a small lake or ship versus small boat), sea state increases (calm to chop to breaking condition with increasing wave height), and visibility decreases (horizon becomes same color as water, water spray or rain on windshield, sunny mid-day versus twilight).

Note. The NCM will advise the P* when the person/equipment is in position on the jungle penetrator. The NCM will perform hoist operations IAW the standard words and phrases IAW unit SOP. The NCM will secure Jungle penetrator or Stokes litter upon completion of the hoisting operation. Should difficulty in maintaining a stable hover occur, the NCM will extend additional cable as slack to preclude inadvertent jerking of the cable.

OVERWATER CONSIDERATIONS:

1. All crewmembers will wear floatation devices IAW AR 95-1.

2. Overwater flight, at any altitude, is characterized by a lack of visual cues; therefore, it has the potential of causing visual illusions. To minimize spatial disorientation, the crew should use radar altitude hold during overwater flight.

3. Be alert to any unannounced changes in the flight profile and be prepared to take immediate corrective actions. The radar altimeter low bug should be set to assist in altitude control.

4. Operations become increasingly more hazardous as references are reduced (open water versus a small lake), water state increases (calm to chop to breaking condition with increasing wave height), and visibility decreases (horizon becomes same color as water, water spray [or rain] on windshield, sunny mid-day versus twilight).

5. Hazards to flight such as harbor lights, buoys, wires, and birds must be considered during overwater flight.

NIGHT OR NIGHT VISION GOGGLES CONSIDERATIONS: Use proper scanning techniques to avoid spatial disorientation.

1. For unaided night flight, the landing light and searchlight should be operational. If a NVG filter is installed, it should be removed.

2.   When NVG are used, hovering with minimum drift is difficult and requires proper scanning techniques and crewmember coordination. If possible, use an area with adequate ground contrast and reference points.

3.   Visual obstacles such as shadows should be treated the same as physical obstacles.

4.   Spatial disorientation can be overwhelming during nighttime overwater operations. If there are visible lights on the horizon or if the shoreline can be seen, the pilot may opt to approach the survivor(s) so the aircraft is pointed toward these references, if the wind permits. If no other references exist, deploy chemlights to assist in maintaining a stable hover.

TRAINING AND EVALUATION REQUIREMENTS:

1.   Training will be conducted in the aircraft.

2.   Evaluation will be conducted in the aircraft.

REFERENCES: Appropriate common references, FM 8-10-6, and TM 55-4240-284-12&P.

TASK 2064

Perform Paradrop Operations

CONDITIONS: In a Mi-17 helicopter, with a jumpmaster and given a designated altitude and appropriate publications.

STANDARDS: Appropriate common standards and the following additions/modifications:

1. RCM.
   a. Properly conduct a thorough crew/passenger briefing.
   b. Ensure the aircraft is properly prepared for the mission.
   c. Maintain airspeed, ±5 knots.
   d. Maintain appropriate altitude, ±100 feet.
   e. Maintain appropriate ground track over the drop zone.
   f. Correctly perform crew coordination actions.

2. NCM.
   a. Properly prepare the aircraft for the mission.
   b. Correctly perform crew coordination actions.

WARNING

Ensure the crew chief and jumpmaster are wearing a safety harness secured to a tie-down ring.

Note. Mi-17 aircraft are only cleared for military free fall parachute operations.

DESCRIPTION:

1. The CE will remove the clam shell doors and ensure the cabin floor is clean and dry. The CE will properly install the static line anchor cable and retriever, as needed, IAW the flight manual. The NCM will ensure the static line anchor cable does not sag more than 6 inches and check the turnbuckle for safety. The NCM will pad and tape every clamp on the cable with cellulose wadding and masking tape. The NCM will rig the troop seats for the mission; adjust the seat backs, as required; and ensure airsick bags are available.

2. The P*/P will thoroughly brief the crewmembers, jumpmaster, and parachutists and ensure the aircraft is properly rigged IAW the flight manual. The P* will maintain altitude, airspeed, and ground track as determined during pre-mission planning. Ground track corrections for wind drift may be received by radio from the drop zone control officer or by intercom from the jumpmaster. The crew will conduct the paradrop according to the procedures covered in the briefing and the references listed below. The PC will verify the jumpmaster or NCM retrieves the static lines as soon as the last parachutist has cleared the aircraft.

Note. If the jumpmaster cannot communicate directly with the P*/P, he or she will communicate with the NCM via hand and arm signals. The NCM will relay necessary information to the P*/P/FE using the intercom.

CAUTION

Ensure the static lines remain secured to the anchor cable until the aircraft has landed. If recovery of the static lines is impossible, execute a running landing to avoid entangling the deployment bags in the rotor system.

> **CAUTION**
>
> When parachutists are equipped with automatic parachute openers and the mission is aborted, ensure the openers are disarmed before beginning the descent.

TRAINING AND EVALUATION REQUIREMENTS:

1. Training will be conducted in the aircraft and/or academically.
2. Evaluation will be conducted in the aircraft and/or academically.

REFERENCES: Appropriate common references and FM 90-26.

TASK 2066

Perform Extended Range Fuel System Operations

CONDITIONS: In a Mi-17 helicopter with internal auxiliary fuel system installed, or academically.

STANDARDS: Appropriate common standards and the following additions/modifications:
1. RCM.
   a. Perform procedures and checks IAW the flight manual, manufacturer's technical manual, or unit SOP.
   b. Operate aircraft within CG/GWT limitations.
2. NCM.
   a. Configure aircraft IAW the flight manual.
   b. Complete all before flight, in-flight, and pre-flight duties IAW the flight manual, manufacturer's technical manual, or unit SOP.
   c. Perform all fuel servicing IAW the flight manual, manufacturer's technical manual, or unit SOP; FM 3-04.111; and FM 10-67-1.

DESCRIPTION: Monitor the left and right fuel quantity indicators and the auxiliary fuel level indicator to ensure the system is operating normally.

TRAINING AND EVALUATION REQUIRMENTS:
1. Training will be conducted in the aircraft and/or academically.
2. Evaluation will be conducted in the aircraft and/or academically.

REFERENCES: Appropriate common references, manufacturer's technical manuals for the systems installed, FM 3-04.111, and FM 10-67-1.

TASK 2081

Operate Night Vision Goggles

CONDITIONS: In a Mi-17 helicopter or a Mi-17 FS.

STANDARDS: Appropriate common standards, and describe and demonstrate correct terminology and usage of the AN/AVS-6 IAW TM 11-5855-263-10.

DESCRIPTION: Perform operational procedures for the AN/AVS-6. These include assembly, preparation for use, operating procedures, and equipment shutdown.

TRAINING AND EVALUATION REQUIREMENTS:
1. Training will be conducted in the aircraft, a Mi-17 FS, or academically.
2. Evaluation will be conducted in the aircraft or academically.

REFERENCES: Appropriate common references and TM 11-5855-263-10.

TASK 2092

Respond to Night Vision Goggle Failure

CONDITIONS: In a Mi-17 helicopter or a Mi-17 FS, with NVG, under NVG conditions, or academically in a classroom environment.

STANDARDS: Appropriate common standards and the following additions/modifications:
1. Correctly identify or describe indications of impending NVG failure.
2. Correctly perform or describe emergency procedures for NVG failure as described in the description.
3. Correctly perform crew coordination actions.

DESCRIPTION: Impending NVG failure is usually indicated by the illumination of the 30-minute, low-voltage warning indicator (AN/AVS-6). Upon indication of NVG failure, perform one of the following procedures:
1. Crew actions. Upon detection of goggle failure, perform the following crew actions:
    a. In flight, P* will announce "goggle failure" with seat position and initiate a climb if obstacle clearance is questionable. The P* will transfer the flight controls if his or her vision is not restored.
    b. The P must be ready to assume aircraft control if the P*announces goggle failure with seat position. If the P announces goggle failure, he or she will perform emergency procedures for NVG failure.
    c. If a NCM announces goggle failure and crew station, he or she will perform emergency procedures for NVG failure.
    d. All crewmembers must be prepared to assume the scan sector assigned to the crewmember whose goggles have failed.
    e. The PC will determine how a crewmember's goggle failure affects the mission and any required deviations.
2. Procedures. Impending NVG failure may be indicated by flickering, flashing, intermittent operation, or by illuminating the low-battery indicator on the visor mount. If the NVG fails, perform the following procedure:
    a. Immediately announce "goggle failure" and crew position.
    b. Attempt to restore NVG power by selecting the alternate battery.
    c. Advise the crew of restored vision or continued failure.
    d. Revise or abort the mission if NVG are not restored.

Note. NVG tube failure is infrequent and usually provides ample warning. Only occasionally will a tube fail completely in a short time. Rarely will both tubes fail at the same time. There is no remedy for in-flight tube failure.

Note. If an NCM experiences goggle failure, the NCM will immediately inform the other crewmembers and attempt to restore vision. The NCM will advise the P*/P of restored vision or of continued failure.

Note. The P*/P should consider aborting or changing the mission if a crewmember's NVG fails and another set is not available. A thorough understanding of options to be used under emergency conditions should be covered in the crew briefing prior to takeoff to avoid any confusion during critical flight maneuvers.

TRAINING AND EVALUATION REQUIREMENTS:
1. Training will be conducted in the aircraft, a Mi-17 FS, or academically.
2. Evaluation will be conducted in the aircraft.

REFERENCES: Appropriate common references and TM 11-5855-263-10.

TASK 2112

Operate Armament Subsystem

> **WARNING**
>
> Observe all safety precautions for uploading ammunition. To prevent accidental firing, do not retract bolt; allow it to go forward if belted ammunition is in feed tray or a live round is in chamber.

CONDITIONS: In a Mi-17 helicopter, with the armament subsystem installed.

Note. This task only applies to flex-fired weapons only.

STANDARDS: Appropriate common standards and the following additions/modifications:

1. Install and pre-flight the armament subsystem IAW the flight manual and subsystem operator's manual.
2. Load and safe the weapon.
3. Acquire and identify the target.
4. Estimate range to target.
5. Engage targets IAW weapon mission briefing, control measures, and ROE.
6. Apply firing techniques.
7. Suppress, neutralize, or destroy as applicable.
8. Describe or perform emergency procedures for misfire, hang-fire, cook-off, runaway gun, and double feeding.
9. Clear and safe the weapon.
10. Enter appropriate information, if required, on DA Form 2408-12, DA Form 2408-13, DA Form 2408-13-1, and aircraft documentation as required.

DESCRIPTION:

1. Crew actions.
   a. NCMs will coordinate with and brief any additional ground support personnel before installing and loading the weapon system. Perform installation and pre-flight inspection of the weapons system.
   b. NCMs will brief all necessary personnel on emergency procedures. They will direct assistance from any additional ground support personnel to aid with installing and loading the weapon. The NCMs also will ensure the proper amount of ammunition is loaded onboard the aircraft IAW the mission briefing.
2. Procedures.
   a. Additional ground support personnel will be briefed as necessary. Perform installation and pre-flight inspection of the weapon, ensuring the gun is safetied to the pintle. Ensure the ejector control bag and ammunition can are installed.
   b. During loading of ammunition, observe all safety precautions while loading. After loading the ammunition, ensure the safety button is in the "S" (safety) position. To initiate the firing sequence, push the safety button to the "F" (fire) position, press the trigger fully, and hold. Low cycle rate of fire of the machine gun allows single-round firing or short bursts.
   c. The trigger must be completely released for each shot. Conduct weapons engagement IAW the mission briefing, ROE, and crew briefing. After acquiring and identifying the target, estimate range and ensure the target is within the weapons field of range and the kill zone is within the weapons effective range.

d.   Use correct firing techniques and ballistic corrections to successfully suppress, neutralize, or destroy the threat, as applicable. Consideration must be given to the visibility of friendly and enemy positions and trying to preclude any undesirable collateral damage or fratricide incidents.

e.   Perform any firing malfunction emergency procedures as required for misfire, hang-fire, cook-off, runaway gun, or double feeding of cartridges. Firing malfunctions and corrective actions must be committed to memory.

f.   Weapons will be cleared and safetied after target engagement. Ensure the safety button is in the "S" position. After completion of the mission, record information as required on DA Form 2408-12, DA Form 2408-13, or DA Form 2408-13-1, and appropriate approved forms as required. Refer to FM 3-04.140 for details on helicopter gunnery qualification.

MULTI-HELICOPTER DOOR GUNNER EMPLOYMENT: Aircrews and door gunners (DGs) in the formation must use effective crew coordination procedures to visually acquire, identify, and engage targets. Both aircraft and passengers are vulnerable to attack during air movement operations and throughout all phases of air assault operations;therefore, it is imperative that DGs respond by delivering direct and indirect fire on these targets. The unit must develop SOPs covering the employment of DGs during formation flights.

NIGHT OR NIGHT VISION GOGGLE CONSIDERATIONS: During night or NVG operations, range estimations will be more difficult which requires using proper scanning techniques. Correct firing techniques and ballistic corrections will be more critical for target suppression or destruction. When wearing NVG during firing, target loss may accrue momentarily due to muzzle blast and the brightness of the tracers.

TRAINING AND EVALUATION REQUIREMENTS:
1.   Training will be conducted in the aircraft and academically.
2.   Evaluation will be conducted in the aircraft.

REFERENCES: Appropriate common references, FM 3-04.140, TM 9-1005-262-13, and TM 9-1005-313-10.

TASK 2125

Perform Pinnacle/Ridgeline Operations

CONDITIONS: In a Mi-17 helicopter or a Mi-17 FS, with the before-landing check completed and hover OGE power available.

STANDARDS: Appropriate common standards and the following additions/modifications:

1. RCM.
    a. Correctly determine power requirements/weight limitations before conducting this task.
    b. Cross major ridgelines at a 45-degree angle.
    c. Correctly determine wind direction for pinnacle landing.
    d. Maintain a constant approach angle.
    e. For transition from terrain flight, align aircraft with landing direction below 100 feet or as appropriate.
    f. Maximum rate of descent during the last 100 feet of a pinnacle approach will not exceed 300 FPM.
    g. Properly clear the aircraft for the landing area.
2. NCM.
    a. Assist in determining the suitability of the landing area for the operation being performed.
    b. Properly clear the aircraft for the landing area.

DESCRIPTION:

1. Crew actions.
    a. Determine power requirements.

    (1) Use current/forecast pressure altitude and temperature to determine power requirements for the conditions at takeoff, cruise, and arrival.

    (2) Before takeoff, analyze winds, obstacles, and density altitude. Perform a hover power check, if required.

    (3) The P* will select a takeoff angle, depending on the wind (demarcation line), density altitude, GWT, and obstacles. After clearing obstacles, accelerate the aircraft to the desired airspeed.

    b. When flying in a valley—

    (1) The aircraft should be flown in the smoother up-flowing air on the lifting side of the valley (windward side).

    (2) Under light winds, the aircraft should be flown closer to the side of the valley. This allows Maximum distance to turn 180 degrees should it become necessary for weather or enemy situation. Additionally, less populated areas are present on the side of the valleys as opposed to the center of the valley. Caution should be used when flying on the leeward side due to potentially significant downdrafts.

    (3) At higher GWTs and pressure altitudes, the Maximum allowable airspeed will decrease. It may be necessary to decrease airspeed to remain within aircraft limitations and prevent blade stall.

    c. Select an approach angle.

    (1) Depending on winds (demarcation line), density altitude, GWT, and obstacles, select an approach angle. An approach angle of 3 degrees or less will minimize the possibility of settling with power.

    (2) During approach, continue to determine the suitability of the intended landing point. The rate of closure may be difficult to determine until the aircraft is close to the landing area. Reduce airspeed to slightly above ETL until the rate of closure can be determined.

    (3) Before reaching the edge of the landing area, reconfirm performance planning and determine if sufficient power will be available.

    (4) Based on performance data, decide whether to continue the approach or make a go-around. If a go-around is required, it should be performed before decelerating below ETL. If the approach is continued, terminate in the landing area to a hover (or to the surface).

(5) After touching down, check aircraft stability as the collective lever is lowered.

Note. Performing this maneuver in certain environments may require hover OGE power. Evaluate each situation for power required versus power available.

Note. A mountain environment is defined IAW the FAR Part 91 for the continental United States. Areas not depicted in FAR Part 91 or host country publications will be identified as mountainous when in an area of steeply sloping terrain, with more than 500-feet elevation relief and terrain elevation more than 5,000 feet above mean sea level).

Note. To successfully operate in small areas, it may be necessary to place the nose of the aircraft over the edge of the landing area. This may cause a loss of important visual references on final approach. In some locations, it may not be possible to lower the forward or aft landing gear on the ground while on/off loading. The description of performing a slope landing in Task 1062 may be used for this type of landing. All crewmembers must assist in providing information on aircraft position in the landing area.

NIGHT OR NIGHT VISION GOGGLE CONSIDERATIONS: More detailed flight planning is required for nighttime flights. When selecting colors for navigational aids (such as maps and kneeboard notes), interior cockpit lighting should be considered.

TRAINING AND EVALUATION REQUIREMENTS:
1. Training may be conducted in the aircraft or a Mi-17 FS.
2. Evaluation may be required in the aircraft or FS at the discretion of the commander.

REFERENCES: Appropriate common references.

TASK 2127

Perform Combat Maneuvering Flight

CONDITIONS: In a Mi-17 or a Mi-17 FS in a training area or tactical environment, with combat maneuvering flight briefing complete.

STANDARDS: Appropriate common standards and the following additions/modifications:
1. Establish entry altitude, ±100 feet.
2. Establish entry airspeed, ±10 KTAS.
3. Maintain the aircraft in trim ±1 ball width.
4. Maintain pitch attitude not to exceed 20 degrees.
5. Maintain roll not to exceed 45 degrees in an accelerated turn.
6. Maintain aircraft within limits and flight envelope.
7. Correctly perform crew coordination actions.
8. Initiate training at altitudes of no lower than 500 feet AHO with a minimum recovery altitude of 200 feet AHO to ensure adequate room to recover.

DESCRIPTION:
1. Crew actions.
    a.   The PC will consider and ensure the crew is aware of the effects of an engine failure during combat maneuvering flight. Airspeed should be maintained between minimum and maximum single-engine airspeed. If an engine failure occurs above or below these airspeeds, engine power will immediately double, associated with possible contingency power application.
    b.   The P* will announce the maneuver to be performed and any deviation from the maneuver. The P* will remain primarily focused outside the aircraft throughout the maneuvers. The primary reference during these maneuvers will be the visible horizon. The P* will make smooth and controlled inputs. Desired pitch and roll angles are best determined by referencing aircraft attitude with the outside horizon. The P* will only momentarily divert focus during critical portions of the maneuver to ensure trim, engine power, and rotor control are maintained. The P* also will announce recovery from the maneuver.
    c.   The P/FE will maintain airspace surveillance and momentarily divert focus during critical portions of the maneuver to ensure trim, engine instruments, maneuver parameters, or aircraft limitations are not exceeded. The P/FE will provide adequate warning to avoid enemy, obstacles, or traffic detected in the flight path and any deviation from the parameters of the maneuver. The P/FE will also announce when his or her attention is focused inside the cockpit; for example, when monitoring airspeed, altitude, and attitude.
2. Procedures.

> *Note.* Performing these maneuvers in certain environments may require hover OGE power: evaluate each situation for power required versus power available.

    a.   Decelerating turn. The decelerating turn is used to rapidly change the direction of the aircraft at low level altitudes while trading airspeed energy to maintain safe operational altitude. The angle of bank, forward airspeed, GWT, and environmental conditions at the initiation of the maneuver will determine the type/amount of deceleration necessary to slow the aircraft to maintain altitude.

    (1)   During flight with lower forward airspeed (typically below maximum rate of climb airspeed), the deceleration will require an increase of collective that results in an increase in blade pitch angle and PTIT. While at airspeeds greater than maximum rate of climb, the airspeed may be traded off while adjusting collective to maintain engine PTIT and blade pitch angle within limits and maintain altitude.

    (2)   Maneuver is typically initiated at airspeeds of 90 to 100 KIAS to effect a direction change while maintaining altitude. For initial training, enter the maneuver at 100 KIAS and the appropriate blade pitch angle. Apply directional cyclic to initiate turn. As the aircraft begins to move about the roll axis,

apply aft cyclic as necessary to maintain altitude by trading airspeed. Apply the pedal as necessary to obtain the appropriate rate of turn. Adjust collective as necessary to maintain altitude and rotor within limits. To recover, apply opposite and forward cyclic while applying opposite pedal and adjusting collective to maintain rotor speed within limits as the rotor system unloads.

Note. For initial training, enter the maneuver at 100 KIAS and appropriate blade pitch angle. Also, do not exceed cruise power setting throughout the maneuver.

b. Break turn. The break turn is used at terrain and cruise flight altitudes to rapidly change the direction of the helicopter while maintaining or gaining airspeed. As altitude allows, this turn also enables a simultaneous three-axis change of position and direction. This maneuver is effective when performing evasive maneuver against small arms, air defense artillery, or to employ weapons.

(1) At cruise altitudes, apply directional cyclic to the initiate turn. The P* will focus his or her attention outside using the horizon as the primary reference for this maneuver. As roll rate and angle increases, the nose may begin to drop. Allow this to occur while maintaining aircraft in trim. Recovery is affected by applying opposite cyclic (roll) when reaching the desired heading. Once the aircraft is wings-level, adjust collective and cyclic to obtain the desired airspeed and altitude.

(2) At terrain flight altitudes, consider desired direction of turn before initiating. Initiate with aft cyclic to ensure adequate obstacle clearance, followed immediately by directional cyclic to initiate turn. Angles of bank are much lower than those utilized during cruise flight since sufficient recovery altitude may not be available.

(3) Maintain trim with pedals. Adjust cyclic as necessary to maintain the pitch attitude as necessary to prevent excessive nose-low attitude to prevent sink-rate build-up.

(4) To recover, apply opposite and forward cyclic.

Note. Maneuver is typically initiated at airspeeds of 60 to 120 KTAS. For initial training, enter the maneuver at 50 KIAS at terrain flight altitudes and 100 KIAS at cruise altitudes. Also, do not exceed cruise power setting throughout the maneuver to simulate operating at maximum power available.

CAUTION

Excessive bank angles at terrain flight altitudes may not allow sufficient recovery time. Airspeed (kinetic energy) may not be available to trade for lift and must be evaluated prior to and during the maneuver. This is aggravated as helicopter GWT and density altitude increase. Do not allow high sink rates to develop, as recovery altitude may not be sufficient. This is aggravated as helicopter GWT and density altitude increase.

c. Dive/dive recovery. This maneuver is used at altitudes above terrain flight to rapidly mask from a threat by placing the aircraft in a dive. This maneuver can be employed when necessary to break contact with enemy fire while maintaining visibility for suppressive fire.

(1) To dive the aircraft as a result of potential enemy contact, apply forward cyclic to obtain the desired dive angle. Adjust the collective to facilitate the rapid descent and maintain the aircraft in trim above 40 KIAS.

(2) Recover at an altitude that will allow sufficient time to arrest the sink rate after collective and cyclic has been applied to recover from the dive. The sink rate may be exacerbated by high GWT.

(3) If the aircraft may have been observed by enemy threat, it may be necessary to turn to an oblique angle of approximately 30 to 45 degrees to evade while minimizing the profile of the aircraft and orienting crew-served weapons for suppressive fire.

Note. Initiate the maneuver for training at no greater than 80 KIAS, not less than 1,000 feet AGL, and recover not less than 200 feet AGL.

Note. During this maneuver, airspeed will increase rapidly. Ensure airspeed does not exceed V_{NE} by initiating a recovery prior to the limit.

NIGHT OR NIGHT VISION GOGGLES CONSIDERATIONS:

1.   Rapid evasive maneuvers will be more hazardous due to division of attention, limited visibility, and aircraft limitations. Be particularly aware of aircraft altitude and three-dimensional position in relation to threat, obstacles, and terrain. Proper sequence and timing is critical in that the P* must announce intentions prior to initiating maneuvers that might cause spatial disorientation.

2.   Select a reference point to maintain orientation on threat or friendly troops to aid in maintaining SA. Reference points may be acquired by selecting a GPS reference point or prominent terrain feature.

3.   As airspeed increases, altitude above the obstacles should also increase. Bank angles should be commensurate with ambient light and altitude above the terrain. High bank angles will result in an inaccurate readout from the radar altimeter; therefore, it is not reliable. Use of NVG will require greater crew coordination to monitor engine instruments, airspeed, trim, and rates of descent information.

Note. While performing combat maneuvering flight, visual contact with other aircraft in the formation may be lost due to maneuvering or reduced visibility. If this occurs, the crewmember should announce loss of visual contact and transmit a call to the other aircraft in the formation.

TRAINING AND EVALUATION REQUIREMENTS:

1.   Training. It is recommended that the Directorate of Evaluation and Standardization (DES) conduct initial training for trainers; however, units are authorized to "self-start" by training and evaluating crewmembers using conditions, standards, and the description as outlined in this task. IPs and FIs will not train or evaluate this task until they have been successfully evaluated by a SP or SI (as appropriate). All other duty designations will be trained and evaluated by a SP/IP or a SI/FI prior to conducting this task. Continuation training may be conducted by qualified crewmembers in the aircraft or a FS .

2.   Evaluations will be conducted in the aircraft.

Note. Crewmembers will ensure that the appropriate authority has authorized this training.

Note. Training combat maneuvering flight with auto-pilot off is not authorized.

REFERENCES: Appropriate common references and FM 3-04.203.

TASK 2169

Perform Aerial Observation

CONDITIONS: In a Mi-17 helicopter, a Mi-17 FS, or academically.

STANDARDS: Appropriate common standards and the following additions/modifications:
1. Detect the target using visual search techniques.
2. Identify the target.
3. Locate the target.
4. Report the target as briefed.

DESCRIPTION:
1. Crew actions.

   a.   All crewmembers are responsible for clearing the aircraft and obstacle avoidance. The P* will maintain aircraft orientation and perform reconnaissance of the assigned sector as duties permit.

   b.   The P will operate the communications systems. When scanning the area, the P should concentrate on avenues of approach while periodically scanning adjoining terrain. The P will select mutually supportive fields of view when working with other aircrews (This will ensure coverage of "dead spaces" that may exist in front of the aircraft). The P will perform reconnaissance of assigned sector and announce when attention is focused inside the cockpit.

   c.   NCMs will assist in clearing the aircraft and provide adequate warning of traffic or obstacles. They will perform observation duties as assigned by the PC and announce when their attention is focused inside the aircraft.

2. Procedures.

   a.   Visual search is the systematic visual coverage of a given area in which every part of the area is observed or scanned. The purpose of a visual search is to detect objects or activities.

   (1) Detection. Detection requires determination that an object or an activity exists.

   (2) Identification. Major factors in identifying a target are size, shape, and type of armament. Targets are classified as friendly or enemy.

   (3) Location. Determining the exact location of targets is the objective of the mission.

   (4) Reporting. Spot reports provide commanders with critical information during the conduct of missions. The method of spot reporting is specified by the requesting agency. Reports of no enemy sightings are frequently just as important as actual enemy sightings.

   b.   The ability of a crewmember to search a given area effectively depends on several factors. In addition to the limitations of the human eye itself, the most important of these factors are altitude, airspeed, terrain and meteorological conditions, and visual cues.

   (1) Altitude. Higher altitudes offer greater visibility with less detail. Lower altitudes are usually used because of survivability considerations.

   (2) Airspeed. Selection of the airspeed is determined by the altitude, the terrain, the threat, and meteorological conditions.

   (3) Terrain and meteorological conditions. The size and details of the area that can be effectively covered largely depend on the type of terrain, such as dense jungle or barren wasteland. The prevailing terrain and meteorological conditions often mask objects and allow only a brief exposure period, especially at NOE altitudes.

   (4) Visual cues. In areas where natural cover and concealment make detection difficult, visual cues may indicate enemy activity. Some of these cues are as follows:

   (a) Color. Colors in nature tend to be subdued. Look for colors that stand out against, d contrast with, natural backdrops.

(b) Texture. Smooth surfaces, such as glass windows or canopies, will shine when reflecting light. Rough surfaces will not.

(c) Shadows. Man-made objects cast distinctive shadows characterized by regular shapes and contours, as opposed to the random patterns occurring naturally.

(d) Trails. Trails leading into an area should be observed for clues as to the type and quantity of traffic, and how recently it passed.

(e) Smoke. Smoke should be observed for color, smell, and volume.

(f) Movement and light. The most easily detectable sign of enemy activity is movement and, at night, light. Movement may include disturbance of foliage, snow, soil, or birds.

(g) Obvious sightings. The enemy is skillful in the art of camouflage. The P*/P must be aware that obvious sightings may be intentional due to high concentrations of antiaircraft weapons.

c. Systematic methods for conducting visual aerial observation include side scan, motive, and stationary techniques. The technique used depends on the altitude flown and the type of terrain.

(1) Side-scan technique. This technique is normally used when the aircraft is operating at an altitude of 100 feet AGL or higher. Over most terrain, the observer systematically—

(a) Looks out approximately 1,000 meters and searches in toward the aircraft.

(b) Looks out one-half the distance (500 meters) and searches in toward the aircraft.

(c) Looks out one-fourth the distance (250 meters) and searches in toward the aircraft.

(d) Motive technique. This technique is used when the aircraft is operating at terrain flight altitudes and at airspeeds of generally 10 KIAS or faster. The entire area on either side of the aircraft is divided into two major sectors: the non-observation sector and the observation work sector. The non-observation sector is the area where the crewmember's field of vision is restricted by the physical configuration of the aircraft. The observation work sector is that portion of the field of vision to which search activity is confined. The observation work sector is subdivided into two smaller sectors: acquisition and recognition.

- The acquisition sector is the forward 45-degree area of the observation work sector. This is the primary area of search.

- The recognition sector is the remainder of the observation work sector. In using the motive technique, the crewmember looks forward of the aircraft and through the center of the acquisition sector for obvious sightings. The crewmember then scans through the acquisition sector, gradually working back toward the aircraft.

(2) Stationary technique. This technique is used at NOE altitudes with the helicopter hovering in a concealed position. When using the stationary technique, the crewmember makes a quick overall search for sightings, unnatural colors, outlines, or movements. The crewmember starts scanning to the immediate front, searching an area approximately 50 meters in depth. The crewmember continues to scan outward from the aircraft, increasing the depth of the search area by overlapping 50-meter intervals until the entire search area has been covered.

NIGHT OR NIGHT VISION GOGGLE CONSIDERATIONS: A thorough crew briefing should be conducted prior to NVG operations. Crew coordination is crucial. Transfer of controls should be covered in detail. When maneuvering the aircraft, the P* must consider obstacles and other aircraft. The P and NCMs should announce when attention is focused inside or outside the cockpit. The P/FE should ensure the P* maintains attention outside the cockpit. All crewmembers must avoid fixation by using proper scanning techniques.

TRAINING AND EVALUATION REQUIREMENTS:

1. Training will be conducted in the aircraft, a Mi-17 FS, or academically.

2. Evaluation will be conducted in the aircraft or academically.

REFERENCES: Appropriate common references and FM 17-95.

This page intentionally left blank.

Chapter 5

Maintenance Test Pilot Tasks

This chapter describes the tasks essential for maintaining maintenance crewmember skills. It defines the task title, number, conditions, and standards by which performance is measured. A description of crew actions, along with training and evaluation requirements is also provided. Tasks described in this chapter are to be performed by a qualified Mi-17 MP IAW AR 95-1. If discrepancies are found between this chapter and appropriate service manuals, TMs, and MTF manuals, the appropriate service manual will take precedence.

5-1. **TASK CONTENTS.**

    a. **Task number.** Each ATM task is identified by a ten-digit systems approach to training number corresponding to the MP tasks listed in table 2-8, page 2-7.

    b. **Task title.** This title identifies a clearly defined and measurable activity. Task titles may be the same in many ATMs, but task content will vary with the airframe.

    c. **Conditions.** The conditions specify the common wartime or training/evaluation conditions under which the MP tasks will be performed.

    d. **Standards.** The standards describe the minimum degree of proficiency or standard of performance to which the task must be accomplished. Standards are based on ideal conditions to which the task must be accomplished. The common standards listed in chapter 4 apply to all tasks listed in this section unless specifically stated otherwise.

    (1) Perform procedures and checks in sequence IAW the appropriate MTF manual.

    (2) Brief the RCM/NCMs on the applicable procedures, warnings, and cautions for the task to be performed.

    (3) Perform aircrew coordination actions IAW the task description and chapter 6.

    (4) Assess and address any malfunctions or discrepancies as they occur and apply appropriate corrective actions or troubleshooting procedures.

    (5) Use the oral callout and confirmation method and announce the initiation and completion of each check.

    (6) The MP must occupy the left seat for Tasks 4022, 4046, 4076, 4086, and 4087. The MP may perform other maintenance tasks from either seat. This restriction does not apply to initial ME training or evaluations.

    (7) The MP/ME will confirm, after moving the engine control levers, that they are back in the detent and the engine is stabilized.

    e. **Description.** The description explains how the elements of the task should be completed to meet the standards. When specific crew actions are required, the task will be broken down as follows:

    (1) Crew actions. These define the portions of a task to be performed by each crewmember to ensure safe, efficient, and effective task execution. When required, MP responsibilities are specified. All tasks in this chapter are to be performed only by qualified MPs, MEs, or student MPs undergoing qualification training as outlined in AR 95-1. The MP is the PC in all situations, except when undergoing training or evaluation by an ME. For all tasks, MP actions and responsibilities are applicable to MEs. When two MEs are conducting training/evaluation together or two MPs are jointly performing TF tasks, the mission brief will designate the aviator assuming PC responsibilities.

    (2) Procedures. This section describes the actions a MP/ME performs or directs the RCM/NCM to perform in order to execute tasks to standard.

Note. Tasks 4010, 4042, 4044, 4093, 4142, 4226, 4236, and 4254 require additional information for the crew. The MP will ensure the crew is familiar with the maneuver (RCM/NCM responsibilities), abort criteria, limitations, and response to associated emergency procedures.

f. **Considerations.** This section defines training, evaluation, and other considerations for task accomplishment under various NVG conditions.

    (1) General.

        (a) Crew selection and aircrew coordination are essential to successful and safe NVG MTFs.

        (b) Tasks may require extra time, altitude, and terrain analysis at night.

        (c) Use of supplemental lighting will aid in identifying switches/position, control positions, and engine control levers.

        (d) Use additional crewmembers to record data as required.

        (e) Use proper scanning techniques to minimize the probability of spatial disorientation.

    (2) Hover checks.

        (a) Select an area with good visual references and room to maneuver during checks.

        (b) Use landing and search lights as required.

    (3) In-flight checks.

        (a) Due to the airspeeds involved while performing several of these checks, select an altitude appropriate for the task.

        (b) Utilize airfields and improved landing environments when available.

g. **Training and evaluation requirements.** Some of the tasks incorporate more than one check from the applicable aircraft MTF manual. For initial MP and RL progressions, all tasks listed in chapter 5 will be evaluated. For APART, the minimum evaluated tasks will be IAW table 2-8, page 2-9. Other tasks/checks will be at the discretion of the ME. Tasks involving dual systems (such as engines) require only one system to be evaluated. Training and evaluation requirements define whether the task will be trained or evaluated in the aircraft, FS, or academic environment. If one or more tasks/checks are not performed to standard, the evaluation will be graded as unsatisfactory; however, when reevaluated, only those unsatisfactory checks must be reevaluated. Evaluations may be conducted in aircraft that are MTF status at the discretion of the ME.

h. **References.** The references are sources of information relating to that particular task. In addition to the common references listed in chapter 4, the following references apply to all MP tasks:

    (1) Aircraft logbook and historical records.

    (2) DA Pam 738-751.

    (3) TM 1-1500-328-23.

    (4) Flight manual.

    (5) Applicable airworthiness directives or messages.

    (6) All approved service manuals and service bulletins.

5-2. **TASK LIST.** The following are the Mi-17 MP tasks.

TASK 4001

Verify Forms and Records

CONDITION: In a Mi-17 helicopter or academically.

STANDARDS: Appropriate common standards.

DESCRIPTION:
1. Crew actions.
   a.   The MP should direct assistance from the P/FE/NCM as necessary.
   b.   The P/FE/NCMs should assist the MP as directed.
2. Procedures.
   a.   Review aircraft forms and records to determine the necessary checks and tasks to be performed. Use additional publications and references as necessary.
   b.   Ensure logbook, phase book, and historical entries are made IAW DA Pam 738-751 or approved forms as required.

TRAINING AND EVALUATION REQUIREMENTS:
1. Training may be conducted in the aircraft or academically.
2. Evaluation will be conducted in the aircraft or academically.

REFERENCES: Appropriate common references.

TASK 4002

Conduct a Maintenance Test Flight

CONDITION: In a Mi-17 helicopter or a Mi-17 FS.

STANDARDS: Appropriate common standards and the following additions/modifications:
1.   Perform the pre-flight inspection IAW the flight manual/CL.
2.   Determine the suitability of the aircraft for flight and the mission to be performed.
3.   Determine the maneuvers, checks, and tasks required during the MTF.
4.   Ensure logbook entries are made IAW DA Pam 738-751 or appropriate aircraft documentation as required.
5.   Perform procedures and checks in sequence IAW the appropriate MTF manual.

DESCRIPTION:
1.   Crew actions.
     a.   The MP should direct assistance from the P/FE/NCM as necessary.
     b.   The P/FE/NCMs should assist the MP as directed.
2.   Procedures.
     a.   Review aircraft forms and records to determine the necessary checks and tasks to be performed. Use additional publications and references as necessary.
     b.   Ensure a thorough pre-flight inspection is conducted with special emphasis on areas or systems where maintenance was performed.
     c.   Verify all test equipment is installed and secured as required.
     d.   The MP will conduct the final walk-around inspection.
     e.   Conduct a thorough crew briefing.

TRAINING AND EVALUATION REQUIREMENTS:
1.   Training may be conducted in the aircraft or a Mi-17 FS.
2.   Evaluation will be conducted in the aircraft or a Mi-17 FS.

REFERENCES: Appropriate common references.

TASK 4004

Perform Interior Check

CONDITION: In a Mi-17 helicopter or a Mi-17 FS.

STANDARDS: Appropriate common standards and the following additions/modifications:

1. Perform procedures and checks in sequence IAW the appropriate MTF manual.

2. Brief the P/FE//NCMs on the applicable procedures, warnings, and cautions for the task to be performed.

DESCRIPTION:

1. Crew actions.
    a. The MP should direct assistance from the P/FE/NCM as necessary.
    b. The P/FE/NCMs should assist the MP as directed.

2. Procedures. Perform the checks IAW the applicable MTF manual.

TRAINING AND EVALUATION REQUIREMENTS:

1. Training may be conducted in the aircraft or a Mi-17 FS.

2. Evaluation will be conducted in the aircraft or a Mi-17 FS.

REFERENCES: Appropriate common references.

TASK 4010

Perform Starting Auxiliary Power Unit Check

CONDITION: In a Mi-17 helicopter or a Mi-17 FS.

STANDARDS: Appropriate common standards and the following additions/modifications:

1. Perform procedures and checks in sequence IAW the appropriate MTF manual.

2. Brief the P/FE//NCMs on the applicable procedures, warnings, and cautions for the task to be performed.

DESCRIPTION:

1. Crew actions.
    a. The MP should direct assistance from the P/FE/NCM as necessary.
    b. The P/FE/NCMs should assist the MP as directed.

2. Procedures. Perform the checks IAW the applicable MTF manual, with the following additional information:
    a. Brief the P/FE/NCM, and any additional ground support personnel as follows:
        (1) APU start abort criteria.
        (2) Monitor and call out times as requested.
    b. Prior to checks, confirm the following:
        (1) Rotor disk clear of APU exhaust.
        (2) Personnel are clear and fire guard posted.

TRAINING AND EVALUATION REQUIREMENTS:

1. Training may be conducted in the aircraft or a Mi-17 FS.

2. Evaluation will be conducted in the aircraft or a Mi-17 FS.

REFERENCES: Appropriate common references.

TASK 4014

Perform Master Warning Check

CONDITION: In a Mi-17 helicopter or a Mi-17 FS.

STANDARDS: Appropriate common standards and the following additions/modifications:

1. Perform procedures and checks in sequence IAW the appropriate MTF manual.

2. Brief the P/FE/NCMs on the applicable procedures, warnings, and cautions for the task to be performed.

DESCRIPTION:

1. Crew actions.
   a. The MP should direct assistance from the P/FE/NCM as necessary.
   b. The P/FE/NCMs should assist the MP as directed.

2. Procedures. Perform the checks IAW the applicable MTF manual.

TRAINING AND EVALUATION REQUIREMENTS:

1. Training may be conducted in the aircraft or a Mi-17 FS.

2. Evaluation will be conducted in the aircraft or a Mi-17 FS.

REFERENCES: Appropriate common references.

TASK 4022

Perform Brake Check

CONDITION: In a Mi-17 helicopter or a Mi-17 FS.

STANDARDS: Appropriate common standards and the following additions/modifications:

1. Perform procedures and checks in sequence IAW the appropriate MTF manual.

2. Brief the P/FE/NCMs on the applicable procedures, warnings, and cautions for the task to be performed.

DESCRIPTION:

1. Crew actions.
   a. The MP should direct assistance from the P/FE/NCM as necessary.
   b. The P/FE/NCMs should assist the MP as directed.

2. Procedures. Perform the checks IAW the applicable MTF manual.

TRAINING AND EVALUATION REQUIREMENTS:

1. Training may be conducted in the aircraft or a Mi-17 FS.

2. Evaluation will be conducted in the aircraft or a Mi-17 FS.

REFERENCES: Appropriate common references.

TASK 4038

Perform Instrument Display System Check

CONDITION: In a Mi-17 helicopter or a Mi-17 FS.

STANDARDS: Appropriate common standards and the following additions/modifications:

1. Perform procedures and checks in sequence IAW the appropriate MTF manual.

2. Brief the P/FE/NCMs on the applicable procedures, warnings, and cautions for the task to be performed.

DESCRIPTION:

1. Crew actions.
   a. The MP should direct assistance from the P/FE/NCM as necessary.
   b. The P/FE/NCMs should assist the MP as directed.

2. Procedures. Perform the checks IAW the applicable MTF manual.

TRAINING AND EVALUATION REQUIREMENTS:

1. Training may be conducted in the aircraft or a Mi-17 FS.

2. Evaluation will be conducted in the aircraft or a Mi-17 FS.

REFERENCES: Appropriate common references.

TASK 4042

Perform Heater and Vent System Check

CONDITION: In a Mi-17 helicopter, a Mi-17 FS, or academically.

STANDARDS: Appropriate common standards and the following additions/modifications:

1. Perform procedures and checks in sequence IAW the appropriate MTF manual.

2. Brief the P/FE/MCNs on the applicable procedures, warnings, and cautions for the task to be performed.

DESCRIPTION:

1. Crew actions.
   a. The MP should direct assistance from the P/FE/NCM as necessary.
   b. The P/FE/NCMs should assist the MP as directed.

2. Procedures. Perform the checks IAW the applicable MTF manual.

> *Note.* Before starting, the heater system should be purged of moisture by operating the blower for 2 minutes. Ensure the heater circuit breaker is in the on/forward position. Check air coming out of all vented areas.

CAUTION

Never start the heater in recirculation mode if cabin temperature exceeds 15 degrees Celsius.

TRAINING AND EVALUATION REQUIREMENTS:

1. Training may be conducted in the aircraft, a Mi-17 FS, or academically.

2. Evaluation will be conducted in the aircraft, a Mi-17 FS, or academically.

REFERENCES: Appropriate common references.

TASK 4043

Perform Windshield Wiper Check

CONDITION: In a Mi-17 helicopter, a Mi-17 FS, or academically.

STANDARDS: Appropriate common standards and the following additions/modifications:

1. Perform procedures and checks in sequence IAW the appropriate MTF manual.

2. Brief the P/FE/NCMs on the applicable procedures, warnings, and cautions for the task to be performed.

DESCRIPTION:

1. Crew actions.
   a. The MP should direct assistance from the P/FE/NCM as necessary.
   b. The P/FE/NCMs should assist the MP as directed.
2. Procedures. Perform the checks IAW the applicable MTF manual.

Note. Ensure windshield remains wet during check to avoid burning out wiper motor and scratching windows.

TRAINING AND EVALUATION REQUIREMENTS:

1. Training may be conducted in the aircraft, a Mi-17 FS, or academically.

2. Evaluation will be conducted in the aircraft, a Mi-17 FS, or academically.

REFERENCES: Appropriate common references.

TASK 4044

Perform Flight Control Hydraulic System Check

CONDITION: In a Mi-17 helicopter or a Mi-17 FS.

STANDARDS: Appropriate common standards and the following additions/modifications:

1. Perform procedures and checks in sequence IAW the appropriate MTF manual.

2. Brief the P/FE/NCMs on the applicable procedures, warnings, and cautions for the task to be performed.

CAUTION

Do not try to center flight controls prior to obtaining hydraulic pressure. Possible damage to bushings, bearings, bellcranks or flight control rods may occur.

DESCRIPTION:

1. Crew actions.
   a.   The MP should direct assistance from the P/FE/NCM as necessary.
   b.   The P/FE/NCMs should assist the MP as directed.

2. Procedures. Perform the checks IAW the applicable MTF manual.

TRAINING AND EVALUATION REQUIREMENTS:

1. Training may be conducted in the aircraft or a Mi-17 FS.

2. Evaluation will be conducted in the aircraft or a Mi-17 FS.

REFERENCES: Appropriate common references.

TASK 4046

Perform Flight Collective Friction Check

CONDITION: In a Mi-17 helicopter or a Mi-17 FS.

STANDARDS: Appropriate common standards and the following additions/modifications:

1.   Perform procedures and checks in sequence IAW the appropriate MTF manual.

2.   Brief the P/FE/NCMs on the applicable procedures, warnings, and cautions for the task to be performed.

DESCRIPTION:

1.   Crew actions.
    a.   The MP should direct assistance from the P/FE/NCM as necessary.
    b.   The P/FE/NCMs should assist the MP as directed.

2.   Procedures. Perform the checks IAW the applicable MTF manual.

TRAINING AND EVALUATION REQUIREMENTS:

1.   Training may be conducted in the aircraft or a Mi-17 FS.

2.   Evaluation will be conducted in the aircraft or a Mi-17 FS.

REFERENCES: Appropriate common references.

TASK 4049

Perform Tail Rotor Pitch Limiter Check

CONDITION: In a Mi-17 helicopter or a Mi-17 FS.

STANDARDS: Appropriate common standards and the following additions/modifications:

1. Perform procedures and checks in sequence IAW the appropriate MTF manual.

2. Brief the P/FE/NCMs on the applicable procedures, warnings, and cautions for the task to be performed.

DESCRIPTION:

1. Crew actions.
    a.   The MP should direct assistance from the P/FE/NCM as necessary.
    b.   The P/FE/NCMs should assist the MP as directed.

2. Procedures. Perform the checks IAW the applicable MTF manual.

TRAINING AND EVALUATION REQUIREMENTS:

1. Training may be conducted in the aircraft or a Mi-17 FS.

2. Evaluation will be conducted in the aircraft or a Mi-17 FS.

REFERENCES: Appropriate common references.

TASK 4064

Perform Beep Trim Check

CONDITION: In a Mi-17 helicopter or a Mi-17 FS.

STANDARDS: Appropriate common standards and the following additions/modifications:

1. Perform procedures and checks in sequence IAW the appropriate MTF manual.

2. Brief the P/FE/NCMs on the applicable procedures, warnings, and cautions for the task to be performed.

DESCRIPTION:

1. Crew actions.
   a. The MP should direct assistance from the P/FE/NCM as necessary.
   b. The P/FE/NCMs should assist the MP as directed.

2. Procedures. Perform the checks IAW the applicable MTF manual.

TRAINING AND EVALUATION REQUIREMENTS:

1. Training may be conducted in the aircraft or a Mi-17 FS.

2. Evaluation will be conducted in the aircraft or a Mi-17 FS.

REFERENCES: Appropriate common references.

TASK 4070

Perform Fuel Quantity Indicator Check

CONDITION: In a Mi-17 helicopter or a Mi-17 FS.

STANDARDS: Appropriate common standards and the following additions/modifications:

1.   Perform procedures and checks in sequence IAW the appropriate MTF manual.

2.   Brief the P/FE/NCMs on the applicable procedures, warnings, and cautions for the task to be performed.

DESCRIPTION:

1.   Crew actions.
    a.   The MP should direct assistance from the P/FE/NCM as necessary.
    b.   The P/FE/NCMs should assist the MP as directed.

2.   Procedures. Perform the checks IAW the applicable MTF manual.

   Note.   Use the outer scale for total fuel readings and the inner scale for individual and auxiliary tanks.

TRAINING AND EVALUATION REQUIREMENTS:

1.   Training may be conducted in the aircraft or a Mi-17 FS.

2.   Evaluation will be conducted in the aircraft or a Mi-17 FS.

REFERENCES: Appropriate common references.

TASK 4072

Perform Barometric Altimeter Check

CONDITION: In a Mi-17 helicopter or a Mi-17 FS.

STANDARDS: Appropriate common standards and the following additions/modifications:

1. Perform procedures and checks in sequence IAW the appropriate MTF manual.

2. Brief the P/FE/NCMs on the applicable procedures, warnings, and cautions for the task to be performed.

DESCRIPTION:

1. Crew actions.
   a. The MP should direct assistance from the P/FE/NCM as necessary.
   b. The P/FE/NCMs should assist the MP as directed.

2. Procedures. Perform the checks IAW the applicable MTF manual.

   Note. Discrepancies greater than ±50 feet require corrective maintenance action. Errors exceeding ±70 feet instrument are not useable for IFR flight.

TRAINING AND EVALUATION REQUIREMENTS:

1. Training may be conducted in the aircraft or a Mi-17 FS.

2. Evaluation will be conducted in the aircraft or a Mi-17 FS.

REFERENCES: Appropriate common references.

TASK 4073

Perform Radar Altimeter Check

CONDITION: In a Mi-17 helicopter or a Mi-17 FS.

STANDARDS: Appropriate common standards and the following additions/modifications:

1.   Perform procedures and checks in sequence IAW the appropriate MTF manual.

2.   Brief the P/FE/NCMs on the applicable procedures, warnings, and cautions for the task to be performed.

DESCRIPTION:

1.   Crew actions.
    a.   The MP should direct assistance from the P/FE/NCM as necessary.
    b.   The P/FE/NCMs should assist the MP as directed.

2.   Procedures. Perform the checks IAW the applicable MTF manual.

Note. Allow 2 minutes for warm up prior to checking.

TRAINING AND EVALUATION REQUIREMENTS:

1.   Training may be conducted in the aircraft or a Mi-17 FS.

2.   Evaluation will be conducted in the aircraft or a Mi-17 FS.

REFERENCES: Appropriate common references.

TASK 4074

Perform Fire Detection System Check

CONDITION: In a Mi-17 helicopter or a Mi-17 FS.

STANDARDS: Appropriate common standards and the following additions/modifications:

1. Perform procedures and checks in sequence IAW the appropriate MTF manual.

2. Brief the P/FE/NCMs on the applicable procedures, warnings, and cautions for the task to be performed.

DESCRIPTION:

1. Crew actions.
   a. The MP should direct assistance from the P/FE/NCMs as necessary.
   b. The P/FE/NCMs should assist the MP as directed.

2. Procedures.
   a. Perform the checks IAW the applicable MTF manual.
   b. Fire extinguisher continuity check.

Note. Ensure the fire extinguishing switch is in the "off" (down) position. Some relays have a 6-second delay prior to light activation when selecting different positions with the fire detector rotary knob.

TRAINING AND EVALUATION REQUIREMENTS:

1. Training may be conducted in the aircraft or a Mi-17 FS.

2. Evaluation will be conducted in the aircraft or a Mi-17 FS.

REFERENCES: Appropriate common references.

TASK 4076

Perform Windshield Anti-Ice Check

CONDITION: In a Mi-17 helicopter, a Mi-17 FS, or academically.

STANDARDS: Appropriate common standards and the following additions/modifications:

1. Perform procedures and checks in sequence IAW the appropriate MTF manual.

2. Brief the P/FE/NCMs on the applicable procedures, warnings, and cautions for the task to be performed.

DESCRIPTION:

1. Crew actions.
   a.   The MP should direct assistance from the P/FE/NCM as necessary.
   b.   The P/FE/NCMs should assist the MP as directed.

2. Procedures. Perform the checks IAW the applicable MTF manual.

Note. **Do not** perform this check at or above 20 degrees Celsius. Ensure the number 2 engine 1919 de-ice valve is closed prior to and after completing check.

TRAINING AND EVALUATION REQUIREMENTS:

1. Training may be conducted in the aircraft, a Mi-17 FS, or academically.

2. Evaluation will be conducted in the aircraft, a Mi-17 FS, or academically.

REFERENCES: Appropriate common references.

TASK 4078

Perform Pitot Heat Systems Check

CONDITION: In a Mi-17 helicopter or a Mi-17 FS.

STANDARDS: Appropriate common standards and the following additions/modifications:

1.   Perform procedures and checks in sequence IAW the appropriate MTF manual.

2.   Brief the P/FE/NCMs on the applicable procedures, warnings, and cautions for the task to be performed.

DESCRIPTION:

1.   Crew actions.
    a.   The MP should direct assistance from the P/FE/NCM as necessary.
    b.   The P/FE/NCMs should assist the MP as directed.

2.   Procedures. Perform the checks IAW the applicable MTF manual.

Note. **Do not** grab pitot tubes after they have been turned on.

TRAINING AND EVALUATION REQUIREMENTS:

1.   Training may be conducted in the aircraft or a Mi-17 FS.

2.   Evaluation will be conducted in the aircraft or a Mi-17 FS.

REFERENCES: Appropriate common references.

TASK 4082

Perform Fuel Boost Pump Check

CONDITION: In a Mi-17 helicopter or a Mi-17 FS.

STANDARDS: Appropriate common standards and the following additions/modifications:

1. Perform procedures and checks in sequence IAW the appropriate MTF manual.

2. Brief the P/FE/NCMs on the applicable procedures, warnings, and cautions for the task to be performed.

DESCRIPTION:

1. Crew actions.
    a.   The MP should direct assistance from the P/FE/NCM as necessary.
    b.   The P/FE/NCMs should assist the MP as directed.

2. Procedures. Perform the checks IAW the applicable MTF manual.

TRAINING AND EVALUATION REQUIREMENTS:

1. Training may be conducted in the aircraft or a Mi-17 FS.

2. Evaluation will be conducted in the aircraft or a Mi-17 FS.

REFERENCES: Appropriate common references.

TASK 4086

Perform Engine Starting System Check

CONDITION: In a Mi-17 helicopter or a Mi-17 FS.

STANDARDS: Appropriate common standards and the following additions/modifications:

1. Perform procedures and checks in sequence IAW the appropriate MTF manual.

2. Brief the P/FE/NCMs on the applicable procedures, warnings, and cautions for the task to be performed.

DESCRIPTION:

1. Crew actions.
   a. The MP should direct assistance from the P/FE/NCM as necessary.
   b. The P/FE/NCMs should assist the MP as directed.

2. Procedures. Perform the checks IAW the applicable MTF manual with the following additional information:
   a. Brief the P/FE/NCMs, and any additional ground support personnel as follows:
      (1) Engine start abort criteria.
      (2) Monitor the flight controls.
      (3) Monitor master fire warning light.
      (4) Call out or note speeds, lights, times, as directed by MP/ME.
   b. Prior to checks, confirm the following:
      (1) The parking brake is set.
      (2) Rotor disk area is clear.
      (3) Ensure fuel fire shutoff valves are "on."
      (4) Personnel are clear and fire guard posted.

TRAINING AND EVALUATION REQUIREMENTS:

1. Training may be conducted in the aircraft or a Mi-17 FS.

2. Evaluation will be conducted in the aircraft or a Mi-17 FS.

REFERENCES: Appropriate common references.

TASK 4087

Perform Engine Abort System Check

CONDITION: In a Mi-17 helicopter or a Mi-17 FS.

STANDARDS: Appropriate common standards and the following additions/modifications:

1.   Perform procedures and checks in sequence IAW the appropriate MTF manual.

2.   Brief the P/FE/NCMs on the applicable procedures, warnings, and cautions for the task to be performed.

DESCRIPTION:

1.   Crew actions.
    a.   The MP should direct assistance from the P/FE/NCM as necessary.
    b.   The P/FE/NCMs should assist the MP as directed.

2.   Procedures. Perform the checks IAW the applicable MTF manual, with the following additional information:
    a.   Brief the P/FE/NCM, and any additional ground support personnel as follows:
        (1)   Perform cold crank.
        (2)   Engine start abort criteria.
        (3)   Monitor the flight controls.
        (4)   Monitor master fire warning light.
        (5)   Call out or note speeds, lights, times, and other information as directed by MP/ME.
    b.   Prior to checks, confirm the following:
        (1)   The parking brake is set.
        (2)   Rotor disk area is clear.
        (3)   Ensure fuel fire shutoff valves are "on."
        (4)   Personnel are clear and fire guard posted.

TRAINING AND EVALUATION REQUIREMENTS:

1.   Training may be conducted in the aircraft or a Mi-17 FS.

2.   Evaluation will be conducted in the aircraft or a Mi-17 FS.

REFERENCES: Appropriate common references.

TASK 4088

Perform Starting Engine Check

CONDITION: In a Mi-17 helicopter or a Mi-17 FS.

STANDARDS: Appropriate common standards and the following additions/modifications:

1. Perform procedures and checks in sequence IAW the appropriate MTF manual.

2. Brief the P/FE/NCMs on the applicable procedures, warnings, and cautions for the task to be performed.

DESCRIPTION:

1. Crew actions.
    a. The MP should direct assistance from the P/FE/NCM as necessary.
    b. The P/FE/NCMs should assist the MP as directed.

2. Procedures. Perform the checks IAW the applicable MTF manual, with the following additional information:
    a. Brief the P/FE/NCMs, and any additional ground support personnel as follows:
        (1) Engine start abort criteria.
        (2) Monitor the flight controls.
        (3) Monitor master fire warning light.
        (4) Call out or note speeds, lights, and times as directed by MP/ME.
    b. Prior to checks, confirm the following:
        (1) The parking brake is set.
        (2) Rotor disk area is clear.
        (3) Ensure fuel fire shutoff valves are "on."
        (4) Personnel are clear and fire guard posted.

TRAINING AND EVALUATION REQUIREMENTS:

1. Training may be conducted in the aircraft or a Mi-17 FS.

2. Evaluation will be conducted in the aircraft or a Mi-17 FS.

REFERENCES: Appropriate common references.

TASK 4090

Perform Engine Run-Up System Check

CONDITION: In a Mi-17 helicopter or a Mi-17 FS.

STANDARDS: Appropriate common standards and the following additions/modifications:

1. Perform procedures and checks in sequence IAW the appropriate MTF manual.

2. Brief the P/FE/NCMs on the applicable procedures, warnings, and cautions for the task to be performed.

DESCRIPTION:

1. Crew actions.
    a. The MP should direct assistance from the P/FE/NCM as necessary.
    b. The P/FE/NCMs should assist the MP as directed.

2. Procedures. Perform the checks IAW the applicable MTF manual.

TRAINING AND EVALUATION REQUIREMENTS:

1. Training may be conducted in the aircraft or a Mi-17 FS.

2. Evaluation will be conducted in the aircraft or a Mi-17 FS.

REFERENCES: Appropriate common references.

TASK 4091

Perform Engine Partial Acceleration Check

CONDITION: In a Mi-17 helicopter or a Mi-17 FS.

STANDARDS: Appropriate common standards and the following additions/modifications:

1. Perform procedures and checks in sequence IAW the appropriate MTF manual.

2. Brief the P/FE/NCMs on the applicable procedures, warnings, and cautions for the task to be performed.

DESCRIPTION:

1. Crew actions.
   a. The MP should direct assistance from the P/FE/NCM as necessary.
   b. The P/FE/NCMs should assist the MP as directed.

2. Procedures. Perform the checks IAW the applicable MTF manual.

TRAINING AND EVALUATION REQUIREMENTS:

1. Training may be conducted in the aircraft or a Mi-17 FS.

2. Evaluation will be conducted in the aircraft or a Mi-17 FS.

REFERENCES: Appropriate common references.

TASK 4092

Perform Engine Dust Cover Protector Check

CONDITION: In a Mi-17 helicopter or a Mi-17 FS.

STANDARDS: Appropriate common standards and the following additions/modifications:

1.   Perform procedures and checks in sequence IAW the appropriate MTF manual.

2.   Brief the P/FE/NCMs on the applicable procedures, warnings, and cautions for the task to be performed.

DESCRIPTION:

1.   Crew actions.
    a.   The MP should direct assistance from the P/FE/NCM as necessary.
    b.   The P/FE/NCMs should assist the MP as directed.

2.   Procedures. Perform the checks IAW the applicable MTF manual.

TRAINING AND EVALUATION REQUIREMENTS:

1.   Training may be conducted in the aircraft or a Mi-17 FS.

2.   Evaluation will be conducted in the aircraft or a Mi-17 FS.

REFERENCES: Appropriate common references.

TASK 4093

Perform Engine Governor Check

CONDITION: In a Mi-17 helicopter or a Mi-17 FS.

STANDARDS: Appropriate common standards and the following additions/modifications:

1. Perform procedures and checks in sequence IAW the appropriate MTF manual.

2. Brief the P/FE/NCMs on the applicable procedures, warnings, and cautions for the task to be performed.

DESCRIPTION:

1. Crew actions.
    a.   The MP should direct assistance from the P/FE/NCM as necessary.
    b.   The P/FE/NCMs should assist the MP as directed.

2. Procedures. Perform the checks IAW the applicable MTF manual.

TRAINING AND EVALUATION REQUIREMENTS:

1. Training may be conducted in the aircraft or a Mi-17 FS.

2. Evaluation will be conducted in the aircraft or a Mi-17 FS.

REFERENCES: Appropriate common references.

TASK 4102

Perform Electrical System Check

CONDITION: In a Mi-17 helicopter or a Mi-17 FS.

STANDARDS: Appropriate common standards and the following additions/modifications:

1. Perform procedures and checks in sequence IAW the appropriate MTF manual.

2. Brief the P/FE/NCMs on the applicable procedures, warnings, and cautions for the task to be performed.

DESCRIPTION:

1. Crew actions.
   a. The MP should direct assistance from the P/FE/NCM as necessary.
   b. The P/FE/NCMs should assist the MP as directed.
2. Procedures. Perform the checks IAW the applicable MTF manual.

TRAINING AND EVALUATION REQUIREMENTS:

1. Training may be conducted in the aircraft or a Mi-17 FS.
2. Evaluation will be conducted in the aircraft or a Mi-17 FS.

REFERENCES: Appropriate common references.

TASK 4112

Perform Taxi Check

CONDITION: In a Mi-17 helicopter or a Mi-17 FS.

STANDARDS: Appropriate common standards and the following additions/modifications:

1. Perform procedures and checks in sequence IAW the appropriate MTF manual.

2. Brief the P/FE/NCMs on the applicable procedures, warnings, and cautions for the task to be performed.

DESCRIPTION:

1. Crew actions.
    a. The MP should direct assistance from the P/FE/NCM as necessary.
    b. The P/FE/NCMs should assist the MP as directed.

2. Procedures. Perform the checks IAW the applicable MTF manual.

TRAINING AND EVALUATION REQUIREMENTS:

1. Training may be conducted in the aircraft or a Mi-17 FS.

2. Evaluation will be conducted in the aircraft or a Mi-17 FS.

REFERENCES: Appropriate common references.

TASK 4119

Perform Systems Instruments Check

CONDITION: In a Mi-17 helicopter or a Mi-17 FS.

STANDARDS: Appropriate common standards and the following additions/modifications:

1.   Perform procedures and checks in sequence IAW the appropriate MTF manual.

2.   Brief the P/FE/NCMs on the applicable procedures, warnings, and cautions for the task to be performed.

DESCRIPTION:

1.   Crew actions.
    a.   The MP should direct assistance from the P/FE/NCM as necessary.
    b.   The P/FE/NCMs should assist the MP as directed.

2.   Procedures. Perform the checks IAW the applicable MTF manual.

TRAINING AND EVALUATION REQUIREMENTS:

1.   Training may be conducted in the aircraft or a Mi-17 FS.

2.   Evaluation will be conducted in the aircraft or a Mi-17 FS.

REFERENCES: Appropriate common references.

TASK 4142

Perform Hover Power/Hover Controllability Check

CONDITION: In a Mi-17 helicopter or a Mi-17 FS.

STANDARDS: Appropriate common standards and the following additions/modifications:

1. Perform procedures and checks in sequence IAW the appropriate MTF manual.

2. Brief the P/FE/NCMs on the applicable procedures, warnings, and cautions for the task to be performed.

DESCRIPTION:

1. Crew actions.
    a. The MP should direct assistance from the P/FE/NCM as necessary.
    b. The P/FE/NCMs should assist the MP as directed.

2. Procedures. Perform the checks IAW the applicable MTF manual.

TRAINING AND EVALUATION REQUIREMENTS:

1. Training may be conducted in the aircraft or a Mi-17 FS.

2. Evaluation will be conducted in the aircraft or a Mi-17 FS.

REFERENCES: Appropriate common references.

TASK 4151

Perform Auto-Pilot Axis Channel Hold Check

CONDITION: In a Mi-17 helicopter or a Mi-17 FS.

STANDARDS: Appropriate common standards and the following additions/modifications:

1.   Perform procedures and checks in sequence IAW the appropriate MTF manual.

2.   Brief the P/FE/NCMs on the applicable procedures, warnings, and cautions for the task to be performed.

DESCRIPTION:

1.   Crew actions.
    a.   The MP should direct assistance from the P/FE/NCM as necessary.
    b.   The P/FE/NCMs should assist the MP as directed.

2.   Procedures. Perform the checks IAW the applicable MTF manual.

TRAINING AND EVALUATION REQUIREMENTS:

1.   Training may be conducted in the aircraft or a Mi-17 FS.

2.   Evaluation will be conducted in the aircraft or a Mi-17 FS.

REFERENCES: Appropriate common references.

TASK 4193

Perform In-Flight Check

CONDITION: In a Mi-17 helicopter or a Mi-17 FS.

STANDARDS: Appropriate common standards and the following additions/modifications:

1. Perform procedures and checks in sequence IAW the appropriate MTF manual.

2. Brief the P/FE/NCMs on the applicable procedures, warnings, and cautions for the task to be performed.

DESCRIPTION:

1. Crew actions.
   a. The MP should direct assistance from the P/FE/NCM as necessary.
   b. The P/FE/NCMs should assist the MP as directed.
2. Procedures. Perform the checks IAW the applicable MTF manual.

TRAINING AND EVALUATION REQUIREMENTS:

1. Training may be conducted in the aircraft or a Mi-17 FS.

2. Evaluation will be conducted in the aircraft or a Mi-17 FS.

REFERENCES: Appropriate common references.

TASK 4194

Perform Flight Instruments Check

CONDITION: In a Mi-17 helicopter or a Mi-17 FS.

STANDARDS: Appropriate common standards and the following additions/modifications:

1. Perform procedures and checks in sequence IAW the appropriate MTF manual.

2. Brief the P/FE/NCMs on the applicable procedures, warnings, and cautions for the task to be performed.

DESCRIPTION:

1. Crew actions.
    a. The MP should direct assistance from the P/FE/NCM as necessary.
    b. The P/FE/NCMs should assist the MP as directed.
2. Procedures. Perform the checks IAW the applicable MTF manual.

TRAINING AND EVALUATION REQUIREMENTS:

1. Training may be conducted in the aircraft or a Mi-17 FS.
2. Evaluation will be conducted in the aircraft or a Mi-17 FS.

REFERENCES: Appropriate common references.

TASK 4204

Perform Compasses, Turn Rate, and Vertical Gyros Checks

CONDITION: In a Mi-17 helicopter or a Mi-17 FS.

STANDARDS: Appropriate common standards and the following additions/modifications:

1. Perform procedures and checks in sequence IAW the appropriate MTF manual.

2. Brief the P/FE/NCMs on the applicable procedures, warnings, and cautions for the task to be performed.

DESCRIPTION:

1. Crew actions.
    a.   The MP should direct assistance from the P/FE/NCM as necessary.
    b.   The P/FE/NCMs should assist the MP as directed.

2. Procedures. Perform the checks IAW the applicable MTF manual.

TRAINING AND EVALUATION REQUIREMENTS:

1. Training may be conducted in the aircraft or a Mi-17 FS.

2. Evaluation will be conducted in the aircraft or a Mi-17 FS.

REFERENCES: Appropriate common references.

TASK 4210

Perform Takeoff and Climb Checks

CONDITION: In a Mi-17 helicopter or a Mi-17 FS.

STANDARDS: Appropriate common standards and the following additions/modifications:

1. Perform procedures and checks in sequence IAW the appropriate MTF manual.

2. Brief the P/FE/NCMs on the applicable procedures, warnings, and cautions for the task to be performed.

DESCRIPTION:

1. Crew actions.
    a.   The MP should direct assistance from the P/FE/NCM as necessary.
    b.   The P/FE/NCMs should assist the MP as directed.

2. Procedures. Perform the checks IAW the applicable MTF manual.

TRAINING AND EVALUATION REQUIREMENTS:

1. Training may be conducted in the aircraft or a Mi-17 FS.

2. Evaluation will be conducted in the aircraft or a Mi-17 FS.

REFERENCES: Appropriate common references.

TASK 4218

Perform In-Flight Controllability Check

CONDITION: In a Mi-17 helicopter or a Mi-17 FS.

STANDARDS: Appropriate common standards and the following additions/modifications:

1.   Perform procedures and checks in sequence IAW the appropriate MTF manual.

2.   Brief the P/FE/NCMs on the applicable procedures, warnings, and cautions for the task to be performed.

DESCRIPTION:

1.   Crew actions.
   a.   The MP should direct assistance from the P/FE/NCM as necessary.
   b.   The P/FE/NCMs should assist the MP as directed.

2.   Procedures. Perform the checks IAW the applicable MTF manual.

TRAINING AND EVALUATION REQUIREMENTS:

1.   Training may be conducted in the aircraft or a Mi-17 FS.

2.   Evaluation will be conducted in the aircraft or a Mi-17 FS.

REFERENCES: Appropriate common references.

TASK 4226

Perform Auto-Pilot In Flight Check

CONDITION: In a Mi-17 helicopter or a Mi-17 FS.

STANDARDS: Appropriate common standards and the following additions/modifications:

1. Perform procedures and checks in sequence IAW the appropriate MTF manual.

2. Brief the P/FE/NCMs on the applicable procedures, warnings, and cautions for the task to be performed.

DESCRIPTION:

1. Crew actions.
   a. The MP should direct assistance from the P/FE/NCM as necessary.
   b. The P/FE/NCMs should assist the MP as directed.

2. Procedures. Perform the checks IAW the applicable MTF manual.

TRAINING AND EVALUATION REQUIREMENTS:

1. Training may be conducted in the aircraft or a Mi-17 FS.

2. Evaluation will be conducted in the aircraft or a Mi-17 FS.

REFERENCES: Appropriate common references.

TASK 4236

Perform Autorotation Revolutions Per Minute Check

CONDITION: In a Mi-17 helicopter or a Mi-17 FS.

STANDARDS: Appropriate common standards and the following additions/modifications:

1. Perform procedures and checks in sequence IAW the appropriate MTF manual.

2. Brief the P/FE/NCMs on the applicable procedures, warnings, and cautions for the task to be performed.

DESCRIPTION:

1. Crew actions.
    a. The MP should direct assistance from the P/FE/NCM as necessary.
    b. The P/FE/NCMs should assist the MP as directed.

2. Procedures. Perform the checks IAW the applicable MTF manual.

TRAINING AND EVALUATION REQUIREMENTS:

1. Training may be conducted in the aircraft or a Mi-17 FS.

2. Evaluation will be conducted in the aircraft or a Mi-17 FS.

REFERENCES: Appropriate common references.

TASK 4252

Perform Vibration Analysis Check

CONDITION: In a Mi-17 helicopter or academically.

STANDARDS: Appropriate common standards and the following additions/modifications:

1. Perform procedures and checks in sequence IAW the appropriate MTF manual.

2. Brief the P/FE/NCMs on the applicable procedures, warnings, and cautions for the task to be performed.

DESCRIPTION:

1. Crew actions.
    a.   The MP should direct assistance from the P/FE/NCM as necessary.
    b.   The P/FE/NCMs should assist the MP as directed.

2. Procedures. Perform the checks IAW the applicable MTF manual.

TRAINING AND EVALUATION REQUIREMENTS:

1. Training may be conducted in the aircraft or academically.

2. Evaluation will be conducted in the aircraft or academically.

REFERENCES: Appropriate common references.

TASK 4254

Perform Velocity Not to Exceed Check

CAUTION

Do not exceed limit as computed on PPC.

CONDITION: In a Mi-17 helicopter or a Mi-17 FS.

STANDARDS: Appropriate common standards and the following additions/modifications:

1.   Perform procedures and checks in sequence IAW the appropriate MTF manual.

2.   Brief the P/FE/NCMs on the applicable procedures, warnings, and cautions for the task to be performed.

DESCRIPTION:

1.   Crew actions.
     a.   The MP should direct assistance from the P/FE/NCM as necessary.
     b.   The P/FE/NCMs should assist the MP as directed.
2.   Procedures. Perform the checks IAW the applicable MTF manual.

TRAINING AND EVALUATION REQUIREMENTS:

1.   Training may be conducted in the aircraft or a Mi-17 FS.

2.   Evaluation will be conducted in the aircraft or a Mi-17 FS.

REFERENCES: Appropriate common references.

TASK 4262

Perform Communication and Navigation Equipment Checks

CONDITION: In a Mi-17 helicopter or a Mi-17 FS.

STANDARDS: Appropriate common standards and the following additions/modifications:

1.   Perform procedures and checks in sequence IAW the appropriate MTF manual.

2.   Brief the P/FE/NCMs on the applicable procedures, warnings, and cautions for the task to be performed.

DESCRIPTION:

1.   Crew actions.
   a.   The MP should direct assistance from the P/FE/NCM as necessary.
   b.   The P/FE/NCMs should assist the MP as directed.

2.   Procedures. Perform the checks IAW the applicable MTF manual.

TRAINING AND EVALUATION REQUIREMENTS:

1.   Training may be conducted in the aircraft or a Mi-17 FS.

2.   Evaluation will be conducted in the aircraft or a Mi-17 FS.

REFERENCES: Appropriate common references.

TASK 4268

Perform Cruise Instrument Check

CONDITION: In a Mi-17 helicopter or a Mi-17 FS.

STANDARDS: Appropriate common standards and the following additions/modifications:

1. Perform procedures and checks in sequence IAW the appropriate MTF manual.

2. Brief the P/FE/NCMs on the applicable procedures, warnings, and cautions for the task to be performed.

DESCRIPTION:

1. Crew actions.
   a. The MP should direct assistance from the P/FE/NCM as necessary.
   b. The P/FE/NCMs should assist the MP as directed.

2. Procedures. Perform the checks IAW the applicable MTF manual.

TRAINING AND EVALUATION REQUIREMENTS:

1. Training may be conducted in the aircraft or a Mi-17 FS.

2. Evaluation will be conducted in the aircraft or a Mi-17 FS.

REFERENCES: Appropriate common references.

TASK 4274

Perform In-Flight Communication/Navigation/Flight Instruments Check

CONDITION: In a Mi-17 helicopter or a Mi-17 FS.

STANDARDS: Appropriate common standards and the following additions/modifications:

1. Perform procedures and checks in sequence IAW the appropriate MTF manual.

2. Brief the P/FE/NCMs on the applicable procedures, warnings, and cautions for the task to be performed.

DESCRIPTION:

1. Crew actions.
    a. The MP should direct assistance from the P/FE/NCM as necessary.
    b. The P/FE/NCMs should assist the MP as directed.

2. Procedures. Perform the checks IAW the applicable MTF manual.

TRAINING AND EVALUATION REQUIREMENTS:

1. Training may be conducted in the aircraft or a Mi-17 FS.

2. Evaluation will be conducted in the aircraft or a Mi-17 FS.

REFERENCES: Appropriate common references.

TASK 4276

Perform Special Equipment and/or Detailed Procedures Checks

CONDITION: In a Mi-17 helicopter or academically.

STANDARDS: Appropriate common standards and the following additions/modifications:

1.   Perform procedures and checks in sequence IAW the appropriate MTF manual.

2.   Brief the P/FE/NCMs on the applicable procedures, warnings, and cautions for the task to be performed.

DESCRIPTION:

1.   Crew actions.
    a.   The MP should direct assistance from the P/FE/NCM as necessary.
    b.   The P/FE/NCMs should assist the MP as directed.

2.   Procedures. Perform the checks IAW the applicable MTF manual.

TRAINING AND EVALUATION REQUIREMENTS:

1.   Training may be conducted in the aircraft or academically.

2.   Evaluation will be conducted in the aircraft or academically.

REFERENCES: Appropriate common references.

TASK 4284

Perform Engine Shutdown Check

CONDITION: In a Mi-17 helicopter or a Mi-17 FS.

STANDARDS: Appropriate common standards and the following additions/modifications:

1. Perform procedures and checks in sequence IAW the appropriate MTF manual.

2. Brief the P/FE/NCMs on the applicable procedures, warnings, and cautions for the task to be performed.

DESCRIPTION:

1. Crew actions.
    a. The MP should direct assistance from the P/FE/NCM as necessary.
    b. The P/FE/NCMs should assist the MP as directed.
2. Procedures. Perform the checks IAW the applicable MTF manual.

TRAINING AND EVALUATION REQUIREMENTS:

1. Training may be conducted in the aircraft or a Mi-17 FS.
2. Evaluation will be conducted in the aircraft or a Mi-17 FS.

REFERENCES: Appropriate common references.

Chapter 6

Aircrew Coordination

This chapter describes the background of aircrew coordination development. It also describes the aircrew coordination principles and objectives, as found in the Army Aircrew Coordination-Enhancement Training Program.

Note. Digitization of crew compartments has expanded and redefined the lines of responsibility for each crewmember. The enhanced ability for either pilot to perform most aircraft/system functions from his or her crew station breaks down the standard delineation of duties and has added capabilities, and potential distractions, in training and combat. This could mean that during an unforeseen event, one pilot may attempt to resolve the situation rather than seeking assistance from or even communicating that action with the other crewmember. It is essential for the PC to brief specific duties prior to stepping into the aircraft. Effective sharing of tasks relies on good crew coordination and information management.

6-1. AIRCREW COORDINATION BACKGROUND AND PLANNING STRATEGY. An analysis of U.S. Army aviation accidents revealed that a significant percentage of aircraft accidents resulted from one or more aircrew coordination errors committed during and even before the flight mission. Often, an accident was the result of a sequence of undetected crew errors that combined to produce a catastrophic result. Additional research showed that even when crews actually avoided potential accidents, these same errors could result in degraded performance that jeopardized mission success. A systematic analysis of these error patterns identified specific areas where crew-level training could reduce the occurrence of such faults and break the chain of errors leading to accidents and poor mission performance.

Aircrew coordination patterns begin with the accomplishment of crew-level pre-mission planning, rehearsal, and AARs. Pre-mission planning includes all preparatory tasks associated with accomplishing the mission. This would include assigning crewmember responsibilities and conducting all required briefings and brief-backs. Pre-mission rehearsal involves the crew collectively visualizing and discussing expected and potential unexpected events for the entire mission. Through this process, all crewmembers discuss and think through contingencies and actions for difficult segments, equipment limitations and failures, or unusual events associated with the mission, and develop strategies to cope with possible contingencies (METT-TC).

Each crewmember must actively participate in the mission planning process to ensure a common understanding of mission intent and operational sequence. The PC prioritizes planning activities so that critical items are addressed within the available planning time. Crewmembers must then mentally rehearse the entire mission by visualizing and discussing potential problems, contingencies, and assigned responsibilities. The PC ensures that crewmembers take advantage of periods of low workload to review or rehearse upcoming flight segments. Crewmembers should continuously review remaining flight segments to identify required adjustments, making certain their planning is consistently ahead of critical lead times.

After a mission or mission segment, the crew should debrief, review, and critique major decisions, their actions, and task performance. This should include identifying options and factors that were omitted from earlier discussion and outline ways to improve crew performance in future missions. Remember, this discussion and critique of crew decisions and actions must remain professional. "Finger pointing" is not the intent and shall be avoided; the emphasis should remain on education with the singular purpose of improving crew and mission performance.

6-2. AIRCREW COORDINATION PRINCIPLES. Broadly defined, aircrew coordination is the cooperative interaction between crewmembers necessary for the safe, efficient, and effective performance of flight tasks. Figure 6-1 provides the essential principles and qualities of aircrew coordination.

Crew Coordination Principles Combine to Produce Coordinated Objectives

Communicate Effectively & Timely

Establish Team Relationships

Sustain a Climate of Ready & Prompt Assistance

Exchange Mission Information

Establish & Maintain Efficient Workloads

Provide Situational Aircraft & Mission Advisories

Cross-Monitor Performance

Manage & Coordinate Actions, Events, & Workloads

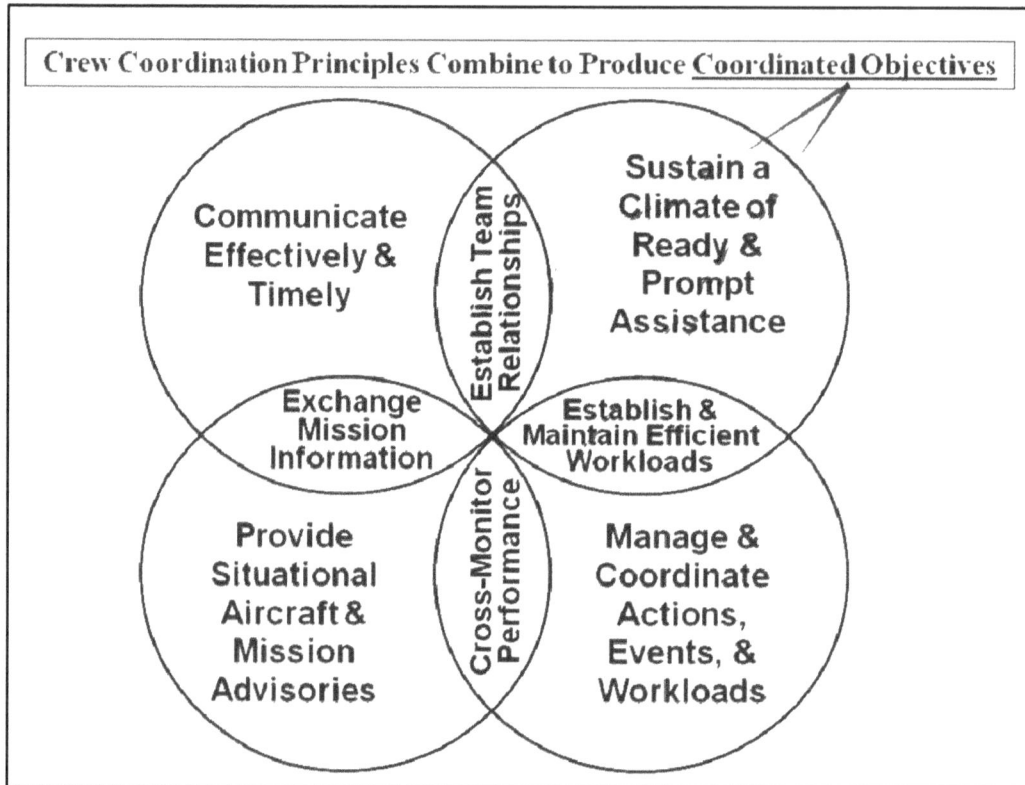

Figure 6-1. Aircrew coordination principles

a. **Communicate Effectively and Timely.** Good team relationships begin with effective communication among crewmembers. Communication is effective when the sender directs, announces, requests, or offers information; the receiver acknowledges the information; and the sender confirms the receipt of information, based on the receiver's acknowledgment or action. This enables the efficient flow and exchange of important **mission information** that keeps a crew on top of any situation that arises.

(1) **Announce and Acknowledge Decisions and Actions.** To ensure effective and well-coordinated actions in the aircraft, all crewmembers must be kept informed and made aware of decisions, expected movements of crew and aircraft, and the unexpected individual actions of others. Each crewmember will announce any actions that may affect the actions of other crewmembers. In turn, communications in the aircraft must include supportive feedback that clearly indicates that crewmembers acknowledge and correctly understand announcements, decisions, or directives of other crewmembers.

(2) **Ensure that statements and directives are clear, timely, relevant, complete, and verified.** These are qualities that must describe the kind of communication that is effective. Considering the fleeting moments of time in a busy aviation environment, only one opportunity may exist to convey critical and **supporting information** before tragedy strikes. This information must be clearly understood, not confusing, and said at the earliest opportunity possible. It must be applicable to the events at hand to support the needs and security of the mission. This information must include all elements needed to make the best decision based on its urgency; and the communication must come with ability of proven confirmation and without redundancy. It must also include the crew's use of standard terminology and feedback techniques that accurately validate information transfer. Emphasis is on the quality of statements associated with navigation, obstacle clearance, instrument readouts, and emergencies. Specific goals include the following:

(a) Crewmembers consistently make the required callouts. Their statements and directives are always timely. Their response to unexpected events is made in a composed, professional manner.

(b)   Crewmembers actively seek feedback when they do not receive acknowledgment from another crewmember. They always acknowledge the understanding of intent and request clarification when necessary.

(3) **Be explicit.** Crewmembers should use clear and concise terms, standard terminology, and phrases that accurately convey critical information. They must avoid using terms that have multiple meanings, such as "right," "back up," or "I have it." Crewmembers must also avoid using indefinite modifiers, such as "Do you see that tree?" or "You are coming in a little fast."

b.   **Sustain a Climate of Ready and Prompt Assistance.** The requirement to maintain a professional atmosphere by all members of the team begins with the team leadership of the PC. However, all crewmembers must equally respect the value of other crewmember's expertise and judgment regardless of rank, duty, or seniority. Every member has a responsibility to maintain SA for mission requirements, flight regulations, operating procedures, and safety. Each crewmember must be willing to practice advocacy and assertiveness should the situation demand a different course of action, as time permits. It is critical to maintain a crew climate that enables opportunity to apply appropriate decision-making techniques for defining the best course of action when problems arise. Courses of action may demand that assistance be directed to other crewmembers or could be voluntary assistance that is offered in a timely manner, depending on time constraints and information available. All crewmembers must remain approachable, especially in critical phases of flight when reaction time is at a premium.

Note. The two-challenge rule allows one crewmember to assume the duties of another crewmember who fails to respond to two consecutive challenges automatically. For example, the P* becomes fixated, confused, task overloaded, or otherwise allows the aircraft to enter an unsafe position or attitude. The P first asks the P* if he or she is aware of the aircraft position or attitude. If the P* does not acknowledge this challenge, the P issues a second challenge. If the P* fails to acknowledge the second challenge, the P assumes control of the aircraft.

c.   **Effectively Manage, Coordinate, and Prioritize Planned Actions, Unexpected Events, and Workload Distribution.** The crew performing as a team should avoid distractions from essential activities while distributing and managing the workloads equally. Both the technical and managerial aspects of coping with normal and unusual situations are important. Proper sequencing and timing guarantees that the actions of one crewmember support and mesh with the actions of the other crewmembers. Responsible effort must be used to ensure that actions and directives are clear, timely, relevant, complete, verified, and coordinated with minimal direction from the PC.

(1) **Direct Assistance.** A crewmember will direct or request assistance when he cannot maintain aircraft control, position, or clearance. A crewmember will also direct assistance when being overloaded with tasks or unable to properly operate or troubleshoot aircraft systems without help from the other crewmembers. The PC ensures that all crew duties and mission responsibilities are clearly assigned and efficiently distributed to prevent the overloading of any crewmember, especially during critical phases of flight. Crewmembers should also watch for workload buildup on others and react quickly to adjust the distribution of task responsibilities.

(2) **Prioritize Actions and Equitably Distribute Workload.** Crewmembers are always able to identify and prioritize competing mission tasks. They never ignore flight safety and other high-priority tasks. They appropriately delay low-priority tasks until those tasks do not compete with tasks that are more critical. Crewmembers consistently avoid nonessential distractions so that these distractions do not affect task performance (for example, sterile cockpit) or the ability to help another crewmember. Crew actions should reflect extensive review of procedures in prior training and pre-mission planning and rehearsal.

d.   **Provide Situational Aircraft Control, Obstacle Avoidance, and Mission Advisories.** Although the P* is responsible for aircraft control, the other crewmembers may need to provide aircraft control information regarding aircraft position (for example, airspeed or altitude), orientation, obstacle avoidance, equipment and personnel status, environmental and battlefield conditions, and changes to mission objectives or evolving situations of the mission (SA). Crewmembers must anticipate and offer supporting information and actions to the decision maker, which is usually the PC or may be the AMC in a mission related situation. Specific goals include the following:

(1) **Situational Awareness.** Crewmembers must anticipate the need to provide information or warnings to the PC or P* during critical phases of the flight or mission. The PC must encourage crewmembers to exercise the freedom to raise issues or offer information about safety or mission related matters. In turn, the crewmembers will provide the required information and warnings in a timely and professional manner. None of this could be accomplished without cross-monitoring performance and crew tasks.

(2) **Mission Changes and Updates.** Crewmembers should routinely update each other while highlighting and acknowledging mission changes. They must take personal responsibility for scanning the entire flight environment, considering their assigned workload and areas of scanning. Each crewmember needs to appropriately adjust individual workload and task priorities with minimal verbal direction from the PC when responding to emergencies and unplanned changes of the mission.

(3) **Offer Assistance.** A crewmember will provide assistance, information, or feedback in response to another crewmember. A crewmember will also offer assistance when he detects errors or sees that another crewmember needs help. In the case where safety or mission performance is at risk, immediate challenge and control measures must be assertively exercised. A crewmember should quickly and professionally inform and assist the other crewmember committing the error. When required, they must effectively implement the two-challenge rule with minimal compromise to flight safety. This means that crewmembers must continually cross-monitor each other's actions and remain capable of detecting errors. Such redundancy is particularly important when crews are tired or overly focused on critical task elements and thus more prone to make errors. Crewmembers must discuss conditions and situations that can compromise SA. These include, but are not limited to, stress, boredom, fatigue, and anger.

6-3. AIRCREW COORDINATION OBJECTIVES. Aircrew coordination principles and objectives originate from and are fundamentally supported by a set of individual, professional skills. Each crewmember is responsible for attaining the leadership skills of effective communication, resource management, decision making, SA, team building, and conflict resolution. When crewmembers are actively using these skills and practicing aircrew coordination principles, results can be seen and measured to determine if the objectives of the aircrew coordination program are being met. The goals of the program have been defined by the following aircrew coordination objectives:

a. **Establish and maintain team relationships.** Establish a positive working relationship that allows the crew to communicate openly, freely, and effectively in order to operate in a concerted manner where a climate of professional assistance is easily found and promptly provided.

b. **Establish and maintain efficient workloads.** Manage and coordinate priorities and execute the mission workload in an effective and efficient manner with the redistribution of task responsibilities as the mission situation changes. Flight duty responsibilities are performed in a timely manner where mission needs are always anticipated.

c. **Exchange mission information.** Establish all levels of crew and mission communications using effective patterns and techniques that allow for the flow of essential data and mission advisories among all crewmembers in a timely and accurate manner.

d. **Cross-monitor performance.** Cross-monitor each other's actions and decisions to ensure workloads and crew actions are performed in a coordinated manner and to standard. Cross-monitoring crewmember performance keeps a crew ready to provide aircraft and mission advisories to each other and helps to reduce the likelihood of errors affecting mission performance and safety.

6-4. STANDARD CREW TERMINOLOGY. To enhance communication and aircrew coordination, crews should use words or phrases that are understood by all participants. They must use clear, concise terms that can be easily understood and complied with in an environment full of distractions. Multiple terms with the same meaning should be avoided. DOD FLIP contains standard terminology for radio communications. The flight manuals contain standard terminology for items of equipment. Table 6-1 provides a list of other standard words and phrases that crewmembers may use.

Table 6-1. Examples of standard words and phrases

Standard word or phrase	Meaning of standard word or phrase
Abort	Terminate a preplanned aircraft maneuver.
Affirmative	Yes.
Arizona	No anti-radiation missiles remaining.
Bandit	An identified enemy aircraft.
Bingo	Fuel state needed for recovery.
Blind	No visual contact of friendly aircraft/ground position. Opposite of VISUAL.
Break	Immediate action command to perform an emergency maneuver to deviate from the present ground track; will be followed by the word "right," "left," "up," or "down."
Call out	Command by the pilot on the controls for a specified procedure to be read from the CL by the other crewmember.
Cease fire	Command to stop firing but continue to track.
Clear	No obstacles present to impede aircraft movement along the intended ground track. Will be preceded by the word "nose," "tail," or "aircraft" and followed by the direction (for example, "left," "right," "slide left," or "slide right"). Also indicates that ground personnel are authorized to approach the aircraft.
Come up/down	Command to change altitude up or down; normally used to control masking and unmasking operations.
Contact	1. Establish communication with...(followed by the name of the element). 2. Sensor contact at the stated position. 3. Acknowledges sighting of a specified reference point (either visually or via sensor). 4. Individual radar return within a GROUP or ARM.
Controls	Refers to aircraft flight controls.
Deadeye	Laser designator system inoperative.
Drifting	An alert of the unintentional or undirected movement of the aircraft; will be followed by the word "right," "left," "backward," or "forward."
Egress	Command to make an emergency exit from the aircraft; will be repeated three times in a row.
Execute	Initiate an action.
Expect	Anticipate further instructions or guidance.
Firing	Announcement that a specific weapon is to be fired.
Fly heading	Command to fly an assigned compass heading. (This term generally used in low-level or contour flight operations.)
Go ahead	Proceed with your message.
Go AJ	Directive to activate antijam communications.
Go plain/red	Directive to discontinue secure operations.
Go secure/green	Directive to activate secure communications.
Hold	Command to maintain present position.

Table 6-1. Examples of standard words and phrases (cont.)

Standard word or phrase	Meaning of standard word or phrase
Hover	Horizontal movement of aircraft perpendicular to its heading; will be followed by the word "left" or "right."
Inside	PRI focus of attention is inside the cockpit for longer than 5 seconds.
Jettison	Command for the emergency or unexpected release of an external (sling) load or stores; when followed by the word "door," will indicate the requirement to perform emergency door removal.
Laser On	Start/acknowledge laser designation.
Lasing	The speaker is firing the laser.
Maintain	Command to continue or keep the same.
Mask/unmask	To conceal aircraft by using available terrain features and to position the aircraft above terrain features.
Mickey	A Have Quick time-synchronized signal.
Monitor	Command to maintain constant watch or observation.
Move aft	Command to hover aft, followed by distance in feet.
Move forward	Command to hover forward, followed by distance in feet.
Negative	Incorrect or permission not granted.
Negative contact	Unable to establish communication with. . . (followed by name of element).
Negative laser	Aircraft has not acquired laser energy.
No joy	Aircrew does not have positive visual contact with the target/bandit/traffic/obstruction/landmark. Opposite of TALLY.
Now	Indicates that an immediate action is required.
Offset (direction)	Maneuver in a specified direction with reference to a target.
Outside	PRI focus of attention is outside the aircraft.
Put me up	Command to place the P* radio transmit selector switch to a designated position; will be followed by radio position numbers on the intercommunication panels (1, 2, 3). Tells the other crewmember to place a frequency in a specific radio.
Release	Command for the planned or expected release of an external (sling) load.
Remington	No ordnance remaining except gun or self-protect ammunition.
Report	Command to notify.
Roger	Message received and understood.
Say again	Repeat your transmission.
Slide	Intentional horizontal movement of an aircraft perpendicular to its heading; will be followed by the word "right" or "left."
Slow down	Command to reduce ground speed.
Speed up	Command to increase ground speed.
Splash	1. (A/S) Weapons impact. 2. (surface-to-surface) Informative call to observer or spotter five seconds prior

Table 6-1. Examples of standard words and phrases (cont.)

Standard word or phrase	Meaning of standard word or phrase
	to estimated time of impact. 3. (air-to-air [A/A]) Target destroyed.
Stand by	Wait; duties of a higher priority are being performed and request cannot be complied with at this time.
Stop	Command to go no further; halt present action.
Strobe	Indicates that the aircraft AN/APR-39 has detected a radar threat; will be followed by a clock direction.
Tally	Sighting of a target, non-friendly aircraft, enemy position, landmark, traffic, or obstruction positively seen or identified; will be followed by a repeat of the word "target," "traffic," or "obstruction" and the clock position. Opposite of No Joy.
Target	An alert that a ground threat has been spotted.
target/object Captured	Specific surface target/object has been acquired and is being tracked with an on-board sensor.
Terminate	Stop laser illumination of a target.
Traffic	Refers to friendly aircraft that present a potential hazard to the current route of flight; will be followed by an approximate clock position and the distance from your aircraft with a reference to altitude (high or low).
Transfer of controls	Positive three-way transfer of the flight controls between the crewmembers (for example, "I have the controls", "You have the controls," and "I have the controls").
Turn	Command to deviate from present ground track; will be followed by words "right" or "left," specific heading in degrees, a bearing ("Turn right 30 degrees), or instructions to follow a well-defined contour ("Follow the draw at 2 O'clock").
Unable	Indicates the inability to comply with a specific instruction or request.
Up on	Indicates PRI radio selected; will be followed by radio position numbers on the intercommunication panels ("Up on 1, up on 3").
Visual	Sighting of a friendly aircraft/ground position. Opposite of BLIND.
Weapons hot/cold/off	Weapon switches are in the ARMED, SAFE, or OFF position.
Wilco	I have received your message, I understand, and I will comply.
Winchester	No ordnance remaining.
Zoom In/Out	Increase/decrease the sensor's focal length. ZOOM IN/OUT is normally followed by "ONE, TWO, THREE, or FOUR": to indicate the number of fields of view (FOVs) to change. (Note: It is recommended only one change in or out at a time be used for the FOV.)

This page intentionally left blank.

Appendix A

Nonrated Crewmember Training and Qualification

A-1.   **CREW CHIEF.** Presently, no formal Army military occupational specialty (MOS) producing school exists for the Mi-17 NCM. Training can be obtained at USAACE; at the unit by completing the aircraft systems, academic, and flight training subjects listed in the following tables; or at an approved training facility.

    a.   **Prerequisites for CE qualification.** U.S. Army service members or DACs must be qualified as a CE in cargo/utility aircraft, possess a current flight physical, and be listed on crewmember orders. Foreign military and civilian personnel must have experience on or be assigned to a Mi-17 unit and possess a current flight physical.

    b.   **Academic qualification training.** The CE must receive sufficient instruction to be knowledgeable in the aircraft manuals, systems, and flight-training subjects listed below. Academic instruction will be IAW this manual. This academic instruction may be completed in any order, but must be completed (to include the examination) and documented in the IATF on DA Form 7122-R (Crew Member Training Record) before flight training. The academic classes are mandatory; however, the hour requirements are based on crewmember retention. Crewmembers must pass the examinations with a grade of at least 70 percent. The required examinations for each subject area are identified in table A-1. Commanders will develop written examinations covering the subject areas listed in this appendix. Each of the following subject areas require a minimum 50-question, open-book examination:

    (1)   The flight manual/systems subjects (to include emergency procedures).

    (2)   Maintenance/service manuals.

    (3)   Academic subjects.

    (4)   Flight training subjects.

Table A-1. Subject area examinations

SYSTEM SUBJECTS	
Aircraft systems, structure, and airframe.	Maintenance forms and records.
Avionics and mission equipment.	Weight and balance.
Flight control hydraulic system.	Electrical system.
Power plant and related systems.	Flight control system.
Auxiliary power unit.	Rotor system.
Transmission and drive systems.	Fuel and oil systems.
Landing gear, wheels, and brake systems.	Environmental systems.
Utility systems.	Prepare aircraft for pre-flight.
Inspection requirements.	Cargo winching and loading.
Aircraft limitations.	Cargo tie down and storage.
Auto-pilot flight control system.	Armaments subsystems.
Aircraft mooring.	Refueling operations.
Required examinations: Maintenance/service manual written examination. System subject written examination. Malfunction analysis (emergency procedures) written examination.	

Table A-1. Subject area examinations (cont.)

ACADEMIC SUBJECTS	
Aeromedical factors.	DA regulations and publications.
Aviation life support equipment.	Passenger briefings.
Unit SOPs and local regulations.	ATP introduction.
Hand and arm signals.	ATM introduction.
Logbook and forms.	In flight duties.
Crew mission briefing.	Confined area and slope operations.
Engine start-through-before takeoff checks.	Aircraft refueling procedures.
External (sling) load operations.	Internal load operations.
Crew coordination training/qualification.	Armament system/operations.
Environmental operations.	Aircraft survivability equipment.
Night mission operations and deployment.	Operating limits and restrictions.
Emergency procedures.	
Required examination: Academic subject written examination	
FLIGHT TRAINING SUBJECTS	
Operating limitations and restrictions.	Pre-flight procedures.
Internal/external (sling) load operations.	In flight duties.
Start and run up procedures.	Radio communication procedures.
Confined area and slope operations.	Before takeoff checks.
Aircraft survivability equipment.	Refueling procedures.
Environmental operations.	Egress procedures.
Required examination: Flight training subject written examination	

c. **Flight training**. The CE will be required to demonstrate proficiency in all individual base tasks listed in table 2-7, page 2-8, and crew coordination and airspace surveillance proficiency. An "X" in the night column of table 2-7 identifies night tasks required for qualification training. Flight hour requirements for aircraft qualification training are based on individual crewmember proficiency. The flight time shown in table A-2 may be used as a guide. Total flight training for aircraft qualification will not be less than 10 hours. Table A-3, page A-3, may be used as a guide for flight time allotted during each training day.

Table A-2. Guide for nonrated crewmember flight training

Flight Instruction	Flying Hours
Base Tasks[1]	9.0
Emergency procedures[2]	2.0
Evaluation[3]	2.0
Total Hours	**13.0**
Notes: 1-A minimum of one hour will be at night. 2-Emergency procedures are required in each mode of flight. 3-The evaluation may be a continual evaluation.	

Table A-3. Guide for flight training sequence

Training Day	1	2	3	4	5
Daily	2.5	2.5	2.5	2.5	3.0E*
Cumulative time	2.5	5.0	7.5	10	13.0
Note: The * denotes night flight and E denotes evaluation. All measurements are in hours					

d. **Documentation.** Upon completion of training, an entry will be made in the remarks section of DA Form 7122-R in the CE's IATF. At the CE's next closeout, training will be documented on the crewmember's DA Form 759 (Individual Flight Record and Flight Certificate-Army), part V, remarks section. A separate entry in the closeout is required for completion of aircraft qualification training.

(1) NVG qualification. NVG qualification will be accomplished IAW paragraph 2-1b, page 2-1.

(2) Refresher training. Refresher training will be accomplished IAW paragraph 2-2, page 2-1.

(3) Mission training. Mission training will be accomplished IAW paragraph 2-3, page 2-4.

(4) Continuation training. Continuation training will be accomplished IAW paragraph 2-4, page 2-6.

(5) CBRNE training. CBRNE training will be accomplished IAW paragraph 2-7, page 2-11.

A-2. **FLIGHT ENGINEER.**

a. **Prerequisites for FE qualification.** U.S. Army service members or DAC must be qualified as a Mi-17 CE, have a minimum of 1 year experience as a crewmember on lift/cargo aircraft, possess a current flight physical, and be listed on crewmember orders. Foreign military and civilian personnel must have qualifications on a Mi-17 and possess a current flight physical.

b. **Initial FE aircraft qualification.** This training is conducted at USAACE and DA-approved training sites IAW a USAACE approved POI. An SP, IP, or SI will conduct initial validation of a crewmember's qualification following this course of instruction and at each new duty station in the aircraft. Additional academic and flight hour requirements are at the discretion of the unit commander.

Table A-4. Guide for flight engineer flight training

Flight Instruction	Flying Hours
Base Tasks [1]	15.0
Emergency procedures [2]	5.0
Evaluation [3]	2.0
Total Hours	**22.0**
Notes: 1-A minimum of one hour will be at night. 2-Emergency procedures are required in each mode of flight. 3-The evaluation may be a continual evaluation	

c. **Documentation.** Upon completion of training, an entry will be made in the remarks section of DA Form 7122-R of the FE's IATF. At the FE's next closeout, training will be documented on the crewmember's DA Form 759, part V, remarks section. A separate entry in the close out is required for completion of aircraft qualification training.

A-3. **FLIGHT INSTRUCTORS AND NON-RATED CREWMEMBER UNIT TRAINERS.**

a. **FI training/qualification.**

(1) **Prerequisites for FI qualification.** U.S. Army service members or DACs must be qualified as a crewmember in cargo/lift aircraft with a minimum of 1 year of experience, possess a current flight physical, and be listed on crewmember orders. Foreign military and civilian personnel must have qualifications on a Mi-17 and possess a current flight physical.

(2) **Initial FI training.** This training is conducted at USAACE, Fort Rucker, Alabama. An SP, IP, or SI will conduct initial validation of a crewmember's qualification following this course of instruction

and at each new duty station in the aircraft. Additional academic and flight hour requirements are at the discretion of the unit commander.

(3) **Documentation.** Upon completion of FI qualification training and evaluation, the SP/IP/SI (as appropriate) will enter the evaluation results on the FI's IATF DA Form 7122-R. Upon completion of a satisfactory evaluation, the DA Form 7120-R (Commander's Task List) will be changed to reflect the new flight duty position and that commander's approval is obtained (initial and date DA Form 7120-R). At the FI's next closeout, training will be documented on the crewmember's DA Form 759, part V, remarks section.

b. **UT qualification training.** The UT was created to lessen the training burden on the FIs/SIs. The UT will only instruct RL-2/RL-1 CEs on certain tasks for which they show an expert knowledge. It was not created to make additional FIs/SIs. Once designated as a UT, he or she may conduct training in the mission/additional tasks he or she is designated to instruct. UTs will not conduct training on FEs or RL-3 CEs, nor will they perform evaluations.

Note. The goal should not be to make all FEs or CEs into UTs in all mission/additional tasks, but rather to give the FEs/CEs the ability to instruct tasks in which they are subject matter experts.

---

(1) **Prerequisites for UT qualification.** The unit commander is responsible for conducting UT qualification IAW this ATM. It is recommended that Active Army and DACs be current Mi-17 RL-1 CEs, possess a current flight physical, and be on crewmember orders.

(2) **Academic training.** Academic training will be conducted at the unit level. The CE must receive sufficient instruction to demonstrate proper method of instruction (MOI) and be knowledgeable in the mission/additional task(s) the CE is designated to instruct. The CE must be able to effectively impart that knowledge to a RL-2 CE.

(3) **Flight training.** The UT will be evaluated on his or her ability to perform, train, and provide MOI for the specific mission/additional tasks in which the UT is designated to instruct. The UT will be required to demonstrate MOI proficiency in designated task(s) and must be able to instruct aircrew coordination and airspace surveillance in those tasks. All flight tasks will be performed to proficiency.

(4) **Documentation.** Upon completion of the UT qualification training and evaluation, the SP/IP/SI/FI (as appropriate) will enter the evaluation results on the CE's IATF DA Form 7122-R. Upon completion of a satisfactory evaluation, the DA Form 7120-R will be changed to reflect the new flight duty position and that commander's approval is obtained (initial and date DA Form 7120-R). At the CE's next closeout, training will be documented on the crewmember's DA Form 759, part V, remarks section.

A-4. **STANDARDIZATION INSTRUCTORS.**

a. **Prerequisites for SI qualification.** U.S. Army service members or DACs must be qualified as a crewmember in cargo/utility aircraft with a minimum of 1 year of experience, qualified as a Mi-17 FI, possess a current flight physical, and be listed on crewmember orders. Foreign military and civilian personnel must have qualifications on a Mi-17 and possess a current flight physical.

b. **Initial SI training.** This training is conducted locally. The SI must be able to supervise and implement the commander's ATP for NCMs and assist the unit SP with the supervision and maintenance of the standardization program. An SP, IP, or SI will conduct initial validation of a crewmember's qualification following this course of instruction and at each new duty station in the aircraft. Additional academic and flight hour requirements are at the discretion of the unit commander.

c. **Documentation.** Upon completion of SI qualification training and evaluation, the SP/IP/SI (as appropriate) will enter the evaluation results on the SI's IATF DA Form 7122-R. Upon completion of a satisfactory evaluation, the DA Form 7120-R will be changed to reflect the new flight duty position and that commander's approval is obtained (initial and date DA Form 7120-R). At the SI's next closeout, training will be documented on the crewmember's DA Form 759, part V, remarks section.

Glossary

AAR	after action review
AGL	above ground level
AHO	above highest obstacle
AIM	aeronautical information manual
ALSE	aviation life support equipment
AMC	air mission commander
ANVIS	aviation night vision imaging system
APART	annual proficiency and readiness test
APU	auxiliary power unit
AR	Army regulation
ASE	aircraft survivability equipment
ASET	aircraft survivability equipment trainer
ATC	air traffic control
ATM	aircrew training manual
ATP	aircrew training program
AWR	airworthiness release
CBAT	computer based aircraft survivability equipment trainer
CBRNE	chemical, biological, radiological, nuclear, and high yeild explosive
CE	crew chief
CG	center of gravity
CI	cockpit indicator
CL	checklist
CSAR	combat search and rescue
CTL	commander's task list
DA	Department of the Army
DA Pam	Department of the Army pamphlet
DAC	Department of the Army civilian
DD	Department of Defense
DG	door gunner
DH	decision height
DOD FLIP	Department of Defense Flight Information Publication
DOT	Department of Transportation
ETA	estimated time of arrival
ETE	estimated time en route
ETL	effective translational lift
ETP	exportable training package
FAA	Federal Aviation Administration
FAC	flight activity category
FAF	final approach fix

FAR	Federal Aviation Regulation
FE	flight engineer
FI	flight engineer instructor
FM	field manual
FPM	feet per minute
FS	flight simulator
GWT	gross weight
GPS	global positioning system
HSI	horizontal situation indicator
IAF	initial approach fix
IATF	individual aircrew training folder
IAW	in accordance with
ICS	internal communication system
IE	instrument flight examiner
IF	intermediate approach fix
IFR	instrument flight rules
IIMC	inadvertent instrument meteorological conditions
IMC	instrument meteorological conditions
IP	instructor pilot
ITO	instrument takeoff
KIAS	knots indicated airspeed
LZ	landing zone
MAP	missed approach point
MAWS	missile approach warning system
ME	maintenance test pilot evaluator
METL	mission-essential task list
METT-TC	mission, enemy, terrain and weather, troops and support available, time available, civil considerations
MO	medical officer
MOC	maintenane operation check
MOI	method of instruction
MOPP	mission-oriented protective posture
MP	maintenance test pilot
MSA	minimum safe altitude
MTF	maintenance test flight
N_1	gas producer (speed)
NAVAID	navigational aid
NCM	nonrated crewmember
N_G	engine speed
NM	nautical mile
NOE	nap-of-the-earth

NOTAM	notice to airmen
N_R	rotor speed
NVD	night vision device
NVG	night vision goggle
NVS	night vision system
OEI	one engine inoperative
OGE	out-of-ground effect
OR	observer
P	pilot not on the controls
P*	pilot on the controls
PA	pressure altitude
PAR	precision approach radar
PC	pilot in command
PFE	proficiency flight evaluation
PMD	preventative maintenance daily
PM-NSRWA	Program Management-Nonstandard Rotary-Wing Aircraft
POI	program of instruction
PPC	performance planning card
PTIT	power turbine inlet temperature
PZ	pickup zone
RCM	rated crewmember
RL	readiness level
ROE	rules of engagement
RPM	revolutions per minute
SFTS	synthetic flight training system
SI	standardization instructor
SM	statute mile
SOI	signal operating instructions
SOP	standard operating procedure
SP	standardization instructor pilot
STANAG	standardization agreement
TC	training circular
TDH	time distance heading
TM	technical manual
TRADOC	Training and Doctrine Command
TSP	training support package
U.S.	United States
USAACE	United States Army Aviation Center of Excellence
UT	unit trainer
VFR	visual flight rules
VMC	visual meteorological conditions

V_{NE}	velocity not to exceed
VSI	vertical speed indicator

blocking	announcement made by the crewmember who intends to block the pedals
bogey	an unidentified aircraft assumed to be enemy
braking	announcement made by the rcm who intends to apply brake pressure
fire	confirmation of illumination of the master fire warning light
hip	north atlantic treaty organization term for the mi-17 helicopter
move back	command to hover aft, followed by a distance in feet
SKED	litter produced by the Skedco Corporation
Stokes	specialized litter developed by Stan Stokes
troops on/off	command for troops to enter/exit the aircraft

This page intentionally left blank.

References

These publications are sources for additional information on the topics in this TC. Most JPs are found at: https://jdeis.js.mil/jdeis/index.jsp. Most Army publications are found online at http://www.apd.army.mil.

SOURCES USED

These are the sources quoted or paraphrased in this publication.

ARMY PUBLICATIONS

AR 95-10. *Department of Defense Notice to Airmen (NOTAM) System.* 3 June 2011.

AR 95-27. *Operational Procedures for Aircraft Carrying Hazardous Materials.* 11 November 1994.

AR 385-10. *The Army Safety Program.* 23 August 2007.

ATTP 3-18.12. *Air Assault Operations.* 1 March 2011.

DA Pam 738-751. *Functional Users Manual for the Army Maintenance Management System-Aviation (TAMMS-A).* 15 March 1999.

FM 1-230. *Meteorology for Army Aviators.* 30 September 1982.

FM 3-04.111. *Aviation Brigades.* 7 December 2007.

FM 3-04.113. *Utility and Cargo Helicopter Operations.* 7 December 2007.

FM 3-04.120. *Air Traffic Services Operations.* 16 February 2007.

FM 3-04.126. *Attack Reconnaissance Helicopter Operations.* 16 February 2007.

FM 3-04.203. *Fundamentals of Flight.* 7 May 2007.

FM 3-04.240. *Instrument Flight for Army Aviators.* 30 April 2007.

FM 3-04.300. *Airfield and Flight Operations Procedures.* 12 August 2008.

FM 3-11. *Multiservice Doctrine for Chemical, Biological, Radiological, and Nuclear Operations.* 1 July 2011.

FM 3-52. *Army Airspace Command and Control in a Combat Zone.* 1 August 2002.

FM 4-20.198. *Multiservice Helicopter Sling Load: Single-Point Load Rigging Procedures.* 20 July 2009.

FM 4-20.199. *Multiservice Helicopter Sling Load: Dual-Point Load Rigging Procedures.* 20 July 2009.

FM 8-10-6. *Medical Evacuation in a Theater of Operations Tactics, Techniques, and Procedures.* 14 April 2000.

FM 10-67-1. *Concepts and Equipment of Petroleum Operations.* 2 April 1998.

FM 17-95. *Cavalry Operations.* 24 December 1996.

FM 21-60. *Visual Signals.* 30 September 1987.

FM 55-450-2. *Army Helicopter Internal Load Operations.* 5 June 1992.

TC 3-04.7. *Army Aviation Maintenance.* 2 February 2010.

TC 3-04.11. *Commander's Aircrew Training Program for Individual, Crew, and Collective Training.* 19 November 2009.

TC 3-04.72. *Aviation Life Support System Management Program.* 15 October 2009.

TC 3-04.93. *Aero Medical Training for Flight Personnel.* 31 August 2009.

TM 1-1500-204-23-1. *Aviation Unit Maintenance (AVUM) and Aviation Intermediate Maintenance (AVIM) Manual for General Aircraft Maintenance (General Maintenance and Practices) Volume I.* 31 July 1992.

TM 1-1500-328-23. *Aeronautical Equipment Maintenance Management Policies and Procedures.* 30 July 1999.

TM 1-1680-377-13&P. *Interactive Electronic Technical Manual (ITEM) for Air Warrior*. 25 May 2012.

TM 4-48.09. *Multiservice Helicopter Sling Load: Basic Operations and Equipment*. 23 July 2012.

TM 9-1005-262-13. *Operators, Aviation Unit and Aviation Intermediate Maintenance Manual for Armament Subsystem, Helicopter, 7.62-mm Machine Gun Mounts: Door Mounted, Lightweight, Model M23 P/N 11691604 (NSN 1005-00-907-0720); Door Mounted, Lightweight, Model M24 P/N 11691606 (NSN 1005-00-763-1404); Ramp Mounted, Lightweight, Model M41 P/N 8436598 (NSN 1005-00-087-2046), and Window Mounted, Lightweight, Model M144 P/N 12011812 (NSN 1005-01-193-4878)*. 29 December 1986.

TM 9-1005-313-10. *Operator's Manual for Machine Gun, 7.62mm, M240 (MSN 1005-01-025-8095) and M240B (1005-01-412-3129), M240C (1005-01-085-4758), M240D (1005-01-418-6995), M240E1 (1005-01-252-4288), M240L (1005-01-549-5837); M240H (1005-01-518-2410); M240N (1005-01-493-1666*. 15 November 2002.

TM 10-1670-201-23. *Organizational and Direct Support Maintenance Manual for General Maintenance of Parachutes and Other Airdrop Equipment*. 30 October 1973.

TM 10-1670-295-23&P. *Unit and Direct Support (DS) Maintenance Manual (Including Repair Parts and Special Tools List) for 10,000lb External Transport Sling Assembly (NSN 1670-01-027-2902) 25,000 lb External Transport Sling Assembly (1670-01-027-2900) 5,000 lb External Cargo Net (1670-01-058-3811) 10,000 lb External Transport Cargo Net (1670-01-058-3810)*. 22 May 1991.

TM 55-1500-342-23. *Joint Service Technical Manual for Aircraft Weight and Balance (NAVAIR 01-1B-50) (USAF TO 1-1B-50) (USCG TO 1-1B-50)*. 30 September 2011.

TM 55-1500-345-23. *Painting and Marking for Army Aircraft*. 12 June 1986.

TM 55-1680-351-10. *Operator's Manual for SRU-21/P Army Vest, Survival (NSN 8465-00-177-4819)(Large) (8465-01-174-2355) (Small)*. 22 April 1987.

DEPARTMENT OF DEFENSE

DOD FLIP. (The DOD FLIP is available from Director, U.S. Army Aeronautical Services Agency, ATTN: MOAS-AI, Cameron Station, Alexandria, VA 22304-5050).

FEDERAL AVIATION ADMINISTRATION

These publications are available from Director, U.S. Army Aeronautical Services Agency, via website: http://www.usaasa.tradoc.army mil/.

Aeronautical Information Manual.

DOT-FAA Order JO 7110.65. *Air Traffic Control*. DOT-FAA Order 8260.42B, *Helicopter Global Positioning System (GPS) Nonprecision Approach Criteria*.

FAA-H-8083-15. *Instrument Flying Handbook 2008*.

FAR Part 91. *General Operating and Flight Rules*. March 1974.

Flight Information Handbook.

STANDARDIZATION AGREEMENT

STANAG 3114 (Edition Eight) /Air Standard 60/16. *Aeromedical Training of Flight Personnel*. 22 October 1986.

NON-STANDARD ROTARY-WING AIRCRAFT TECHNICAL AVIATION LIBRARY

These publications are available through the Program Management-Nonstandard Rotary-Wing Aircraft office at https://upw.jtdi mil/and directed for use with the Mi-17 series aircraft.

Mi-17-1V Helicopter Flight Manual. 5 October 2000.

Mi-17-1V Helicopter Maintenance Manual–General with Change 1. 20 April 2011.

Mi-17-1V Helicopter Maintenance Manual–Airframe. 29 April 2008.

Mi-17-1V Helicopter Maintenance Manual–Power Plant. 09 November 2007.

Mi-17-1V Helicopter Maintenance Manual–Helicopter Systems. 5 October 2000.

Mi-17-1V Helicopter Maintenance Manual–Helicopter Equipment. 13 May 2009.

Mi-17-1V Helicopter Maintenance Manual–Pyrotechnic Devices and Aerial Delivery Equipment. 9 November 2007.

Mi-17-1V Fire Extinguishers Maintenance Schedule. 25 April 1977.

Mi-17-1V Helicopter Maintenance Schedule–Airframe, Helicopter Systems, Power Plant. 9 July 2009.

Mi-17-1V Helicopter Maintenance Schedule–Avionics. 5 October 2000.

Mi-17-1V Helicopter Maintenance Schedule–Helicopter Equipment. 5 October 2000.

Mi-17-1V Helicopter Maintenance Schedule–Pyrotechnic Devices And Aerial Delivery Equipment. 5 October 2000.

Mi-17-1V OEM/Vendor Maintenance Manuals.

Mi-17-B Helicopter Standard Specification.

DOCUMENTS NEEDED

These documents must be available to the intended users of this publication.

ARMY PUBLICATIONS

AR 40-8. *Temporary Flying Restrictions Due to Exogenous Factors.* 16 May 2007.

AR 70-62. *Airworthiness Qualification of Aircraft Systems.* 21 May 2007.

AR 95-1. *Flight Regulations.* 12 November 2008.

AR 95-2. *Airspace, Airfields/Heliports, Flight Activities, Air Traffic Control, and Navigational Aids (RAR 001, 10/16/2008).* 10 April 2007.

AR 95-20. *Contractor's Flight and Ground Operations (DMCA INST 8210.0; AFI 10-220; NAVAIRINST 3710.1F; COMDTINST M13020.3).* 1 March 2007.

AR 190-11. *Physical Security of Arms, Ammunition and Explosives (RAR 001, 06/28/2011).* 15 November 2006.

AR 190-51. *Security of Unclassified Army Property (Sensitive and Nonsensitive).* 30 September 1993.

AR 600-105. *Aviation Service of Rated Army Officers.* 22 June 2010.

AR 600-106. *Flying Status for Nonrated Army Aviation Personnel.* 8 December 1998.

FM 3-22.68. *Crew Served Weapons.* 21 July 2006.

FM 3-04.140. *Helicopter Gunnery.* 14 July 2003.

FM 3-04.203. *Fundamentals of Flight.* 7 May 2007.

FM 3-21.220. *Static Line Parachuting Techniques and Tactics (MCWP 3-15.4; AFMAN 11-420; NAVSEA SS400-AF-MMO-010).* 23 September 2003.

FM 90-26. *Airborne Operations.* 18 December 1990.

TC 3-04.11. *Commander's Aircrew Training Program for Individual, Crew, and Collective Training.* 19 November 2009.

DEPARTMENT OF THE ARMY FORMS

DA Form 759. *Individual Flight Record and Flight Certificate-Army.*

DA Form 2028. *Recommended Changes to Publications and Blank Forms.*

DA Form 2408-12. *Army Aviator's Flight Record.*

DA Form 2408-13. *Aircraft Status Information Record.*

DA Form 2408-13-1. *Aircraft Inspection and Maintenance Record.*

DA Form 4186. *Medical Recommendation for Flying Duty.*

DA Form 5484. *Mission Schedule/Brief.*

DA Form 5701-17. *Mi-17 Performance Planning Card.*

DA Form 7120-R. *Commander's Task List.*

DA Form 7122-R. *Crew Member Training Record.*

DA Form 7382. *Sling Load Inspection Record.*

DEPARTMENT OF DEFENSE FORMS

DD Form 365-4. *Weight and Balance Clearance Form F-Transport/Tactical.*

READINGS RECOMMENDED

These sources contain relevant supplemental information.

ADP 7-0. *Training Units and Developing Leaders.* 23 August 2012.

Index

A

air traffic control (ATC), 4-3

aircrew training program (ATP), vii, 1-2, 2-1, 3-1, 3-2, 3-3, 4-1, A-4

annual proficiency and readiness test (APART), 1-1, 2-9, 3-3

automatic flight control system (AFCS), 4-2

aviation life support equipment (ALSE), 3-4

aviator's night vision imaging system (ANVIS), 1-1

C

chemical, biological, radiological, and nuclear (CBRN), 2-10

commander's task list (CTL), 1-1, 2-6, 2-8, 2-9, 2-10, 3-2, 3-3, 3-7

crew coordination, 3-1, 3-4, 4-1, 4-2, 4-3, A-2, A-4

currency
    aircraft, 2-10
    NVG, 2-10

E

evaluation
    academic, 3-2, 3-3
    crewmember, 3-2
    debriefing, 3-7
    flight, 3-2, 3-7
    guidelines, 2-9
    introduction, 3-3
    principles, 3-1
    sequence, 3-3

exportable training package (ETP), 2-1

F

flight activity category (FAC), 2-5, 2-8
    1, 2-5
    2, 2-5
    3, 2-5, 2-10

G

grading considerations, 3-1

I

inadvertent instrument meteorological conditions (IIMC), 3-4

individual aircrew training folder (IATF), 3-2, 3-3, A-1, A-3, A-4

instrument flight rule (IFR), 3-3, 3-4

instrument meteorological conditions (IMC), 3-7

M

maintenance test flight (MTF), 2-8

minimum flight hours, 2-1

mission essential task list (METL), 1-1, 1-2, 2-2, 2-4, 2-9, 2-10

N

night vision device (NVD), 1-1

P

personnel terminology, 1-1

primary flight examiner (PFE), 1-2

Q

qualification
    aircraft, 2-1
    NVG, 2-1

R

readiness level (RL), 2-1, 2-2, 2-5, 2-10
    2, 1-2
    3, 2-1

requirements
    annual, 2-5
    semiannual, 2-5

S

symbol usage, 1-1

synthetic flight training system (SFTS), 2-5

T

tasks
    base, 2-9
    maintenance test pilot, 2-9
    mission, 2-9
    performance, 2-9
    technical, 2-9

training
    academic, 2-1, 2-2, 2-5, A-4
    flight, 2-1, 2-2, 2-5, A-1, A-2, A-4
    refresher, 2-2

training support package (TSP), 2-1

V

visual flight rules (VFR), 3-4

W

word distinctions, 1-1

This page intentionally left blank.

By order of the Secretary of the Army:

RAYMOND T. ODIERNO
General, United States Army
Chief of Staff

Official:

JOYCE E. MORROW
Administrative Assistant to the
Secretary of the Army
1211801

DISTRIBUTION:

Active Army, Army National Guard, and United States Army Reserve: Not to be distributed; electronic media only.